UNDERWATER ADVENTURE

Suddenly both hands shot out towards the moray's neck. But the moray moved faster and the powerful jaws closed on Hal's wrist. The sharp teeth bit painfully into his flesh. A thin trickle of Hal's blood oozed from the moray's mouth.

Attracted by the blood, the shark again pushed its great face into the cave entrance, shutting off the light. Hal tried to pull his arm away, but the teeth only sank deeper.

If he struggled he would lose his arm. He must be patient. If this great eel followed the habits of other eels, sooner or later it would relax its grip in order to take a better hold. In that instant, he might wrench his arm away.

But it was an agony to be patient under such circumstances. To make matters worse, the shark, excited by the smell of blood, began battering the entrance with his armour-plated head. Chunks of coral fell and the opening grew wider.

by the same author

AMAZON ADVENTURE
SOUTH SEA ADVENTURE
VOLCANO ADVENTURE
WHALE ADVENTURE
AFRICAN ADVENTURE
ELEPHANT ADVENTURE
SAFARI ADVENTURE
LION ADVENTURE
GORILLA ADVENTURE
DIVING ADVENTURE
CANNIBAL ADVENTURE
TIGER ADVENTURE
ARCTIC ADVENTURE

UNDERWATER ADVENTURE

by

WILLARD PRICE

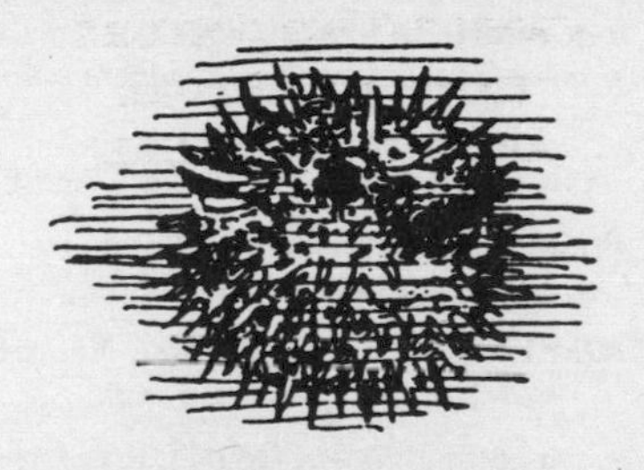

Illustrated from drawings by

PAT MARRIOTT

RED FOX

A Red Fox Book
Published by Random House Children's Books
61-63 Uxbridge Road, London W5 5SA

A division of Random House UK Ltd

London Melbourne Sydney Auckland
Johannesburg and agencies throughout the world

First published by Jonathan Cape 1955

First paperback edition Knight Books 1971

Red Fox edition 1993

15 17 19 20 18 16 14

Set in Baskerville
Typeset by JH Graphics Ltd, Reading, Berkshire

Printed and bound in Great Britain by
Cox & Wyman Ltd, Reading, Berkshire

RANDOM HOUSE UK Limited Reg. No. 954009
www.kidsatrandomhouse.co.uk

Papers used by Random House UK Limited
are natural, recyclable products made from wood grown in sustainable forests. The manufacturing processes conform to the environmental regulations of the country of origin

ISBN 0 09 918231 9

Contents

Illustrations

FOR

KEN

UNDERWATER ADVENTURE

1
Mask and Snorkel

The gallant little ship, *Lively Lady*, rode at anchor in the lagoon of Truk, paradise atoll of the South Seas.

On every side rose high islands, clothed from base to summit with coconut palms, breadfruit and mango trees, and blazing bougainvillea.

There are two hundred and fifty islands in Truk lagoon. The lagoon is huge, forty miles across, a veritable lake in the ocean. It is girdled by a low coral reef. Four breaks in the reef allow ships to pass into and out of the lagoon.

Below the *Lively Lady* the water was so clear that Hal and Roger, looking over the rail, could plainly see the gorgeous coral gardens on the bottom forty feet down.

The two brothers, Hal in his late teens, Roger in his early teens, had been allowed a year off from school to make certain expeditions for their father, John Hunt, the famous animal-collector. They had spent the summer in journeys through the Amazon jungle and the Pacific, collecting wild animals and great fish for the zoos and circuses – a story already told in *Amazon Adventure* and *South Sea Adventure*.

Now they were to go beneath the sea. Their father, following his plan to give them a practical education in natural history, had lent them to the Oceanographic Institute.

The Institute had outfitted the *Lively Lady* with diving bell, aqualungs, submarine cameras and other equipment for deep-sea operations to study the habits of great fish and capture specimens. Bob Blake, doctor of science, had been assigned by the Institute to direct the work.

Dr Blake looked more like a lifeguard than a scientist. His skin, where it was not covered by the yellow bathing trunks, was tanned a deep mahogany. His shoulders were broad and his big chest and tough arm muscles suggested a powerful swimmer. His face was intelligent, but just now it was screwed up into a gloomy scowl. He sat on a hatch and studied the two boys at the rail.

'Why, oh why . . .' he was thinking, 'why did I have to draw these two amateurs? What do they know about deep-sea diving? They've probably never been deeper than the bottom of a bathtub.'

His eye followed Hal from head to foot.

'A lot of man there,' he had to admit to himself.

'Half my age, and bigger than I am. A steady, sensible fellow. And his kid brother is a likeable youngster. But that doesn't make them deep-sea divers. Well, if I have to play kindergarten teacher, I may as well get started.'

He called the boys. 'Time for our first diving lesson.'

They came eagerly and joined him on the hatch. Captain Ike, owner and skipper of the vessel, edged closer. Omo, the young Polynesian seaman, dangling above in a bosun's chair, rested from his job of sandpapering the masts and listened.

'You boys know,' Dr Blake began, 'that seventy per cent of the earth's surface is water. Most of the land has been explored. But we have only begun to explore the water. The world beneath the sea is yet to be discovered. The great explorations of the next hundred years will be in the ocean depths.

'Scientists have tried to learn what goes on down there by letting down nets, then studying the fish and weeds that come up in the nets. That's a very poor way of going at it. A much better way is to go down and see for ourselves. But this was not easy in the old-fashioned diving suit because it was so clumsy and dangerous.

'Recently there have been some marvellous inventions that make it possible for us to go down into the sea and feel at home there. One is the snorkel. One is the aqualung. One is the diving bell. One is the undersea sled.

'We have all of these on board. What I want you to do is to become familiar with their use so that you can help me study deep-sea life, take underwater photographs, and capture specimens. I know you have had some training in zoology in your father's animal business. And they tell me you made a fine record on your Amazon and Pacific expeditions.'

Hal and Roger glowed with pleasure. Dr Blake took the wind out of their sails with his next remark.

'But all that won't help you much. The main thing in this job is to be able to dive. How much diving experience have you had?'

'Mighty little,' Hal said truthfully.

'I thought so. Now the first thing I want you to do is to jump over the side and let me see how far down you can go. If you have sharp pain in the ear-drums, come up at once. You'll be lucky if you swim ten feet deep on the first try.'

Roger leaped up on the rail. He would show this doubting professor. He prided himself on his ability to make a clean high jack-knife and a deep plunge. But Blake stopped him.

'Hold it! Not that way. No plunge. That would scare the fish.'

'How else?' asked bewildered Roger.

'Go in the way an old woman would. Let yourself down into the water, gently, without making a splash.'

Hal and Roger eased themselves over the rail and into the lagoon without raising a ripple. Then they upended and swam down.

It was Dr Blake's turn to be bewildered. He had expected the boys to go only a few feet deep and come up gasping and sputtering. Instead, with long even strokes, they swam down, down, down. Ten feet, twenty feet, thirty, and down to the forty-foot bottom.

Their brown friend, Omo, watched their performance proudly and enjoyed the look of surprise on the professor's face. Omo himself was not surprised, for he well remembered the experience his companions had gained in diving for pearls on their previous expedition.

Dr Blake threw a line over the side. The two boys rose swiftly to the surface, shot out like porpoises, seized the line and swarmed up it to the deck.

There they stretched out in the sun. Their breathing came in gasps and their faces showed the strain of the dive. They waited for Dr Blake to say something.

But their instructor did not believe in distributing praise with a lavish hand.

'Not bad,' he said, 'for beginners. But you'll do better if you bounce first.'

'Bounce?' queried Hal.

'This way.'

Blake let himself down smoothly over the rail, then swam slowly to a point where the lagoon was about sixty feet deep. He sank until only a bit of his brown hair could be seen. With a sudden thrust of arms and legs, he shot up out of the water waist-high and sank, still in an upright position, to a depth of eight or ten feet. Then he doubled and swam down, so swiftly that it seemed he could not have gone halfway to the bottom before he popped from the water holding a sea fan plucked from the coral floor.

Hal and Roger realized their good luck in having a diving master who could do as well as teach.

Blake clambered aboard. His breathing was normal and he looked as calm as if he had dived six feet instead of sixty.

'Lesson number two,' he said. 'Have you ever used a snorkel?'

The boys shook their heads. Blake opened a case and took out face masks, swim fins and snorkels.

'Then you're in for a treat,' he said. 'Put these on.'

The boys were familiar with masks and fins and easily slipped them in place.

But the snorkel! They examined this device with curiosity. It was a plastic tube about two feet long. It looked like a snake bent up at one end and down at the other. On one end was a mouthpiece.

'Put that in your mouth. The rubber flanges go behind your lips and you grip those rubber nubbins between your teeth. Then it doesn't matter if your head is under water – you can breathe – as long as the other end of the tube is above the surface.'

'But,' objected Roger, 'suppose the sea is rough – waves splashing against the snorkel – won't you take in water instead of air?'

'See the ping-pong ball in the little cage at the top end?' said Blake. 'When a wave comes it throws the ball up, closing the opening. No water comes through the tube. When the wave falls away, the ball drops and you breathe again. You'll find with a little practice that you won't even notice these interruptions.'

'How long can you keep your head under water with one of these?'

'Why, all day if you like. It's as easy as breathing. The only difference is you breathe through your mouth instead of your nose. That's no trick. Lots of snorers do it every night.'

'Why do they call the thing a snorkel?' Roger wanted to know.

'During World War II one type of German U-boat had a *schnorkel* — a tube to bring down air to the submarine. Our word is the same, only simplified.'

The diving master put on a mask and fins, and selected a snorkel. 'I'll show you how it's used.' He slipped the rubber mouthpiece behind his lips. He went over the rail and lay in the water, face down, almost submerged, only the back of his head above the surface. The end of the snorkel rose out of the sea like the head of a sea serpent.

When a ripple splashed against the serpent's head, the ball it held in its teeth popped up, and momentarily closed the opening. Blake swam about lazily, viewing the coral gardens below through the window of his mask. Then he dived. As the upper end of the snorkel sank into the sea, water pressure forced the ball into the end of the breathing tube. When the diver rose and the snorkel emerged into the air, the ball fell away and the diver could breathe again.

For fifteen minutes Blake swam about, just beneath the surface, diving occasionally, and never once took his face out of the water.

'It's as easy as lying in bed,' he said as he climbed on board. 'Try it. One at a time.'

'Me first,' said Roger eagerly.

He clamped lips and teeth over the snorkel mouthpiece and slid into the sea.

He floated face down as Blake had done. But old habits were too strong. He held his breath as he had always had to hold his breath under water. Then he tossed his head out of the water to take air, but when he opened his mouth to breathe, out fell the snorkel. He could hear Blake scolding him. He replaced the mouthpiece in his mouth. He reminded himself that with the snorkel you could actually breathe under water.

He deliberately sank his face into the lagoon and kept it there. He tried to breathe – through his nose. Since the face-plate that covered his eyes covered his nose also, inhaling gave him no air but merely tightened the mask on his face.

Of course – he must breathe through the mouth. He tried it, and the air flowed easily into his lungs. He exhaled, inhaled again, exhaled . . . why, there was nothing to it. All you had to do was forget that you were in the water. Forget that the sea was your enemy. Make it your friend. Rest in its arms.

He relaxed. He was breathing regularly now, although it still seemed strange to be inhaling fresh, dry air under water. Although he was a good swimmer, he had always had to fight the sea. Fight to breathe, fight to keep water out of the nose, fight to avoid swallowing a bucketful, fight to stay up, fight to swim down, fight to force a passage through the waves.

Now there was no fight. The tension left his arms and legs. He lay in the warm tropical water as in a feather bed. He knew there were waves above him because he had seen them before he went down. But now they merely washed over him and he felt nothing but a gentle swinging motion. Now and then a wave would bury the snorkel and the ping-pong ball would shut off his air, but it would be only an instant before he could breathe again. Soon he did not even notice these slight breaks.

He thought of how different this was from floating on the surface with your face up. Then you had no peace. You had to be on guard every moment lest a wave slop over your face and bring you up choking and gagging. You couldn't look down. You could see nothing but the bare sky. You had to keep your lungs well filled. If your feet were heavy, as his were, you must struggle to keep them up.

Lying face down, he had none of these troubles. Why his feet didn't sink he could not understand. Perhaps it was because his head was completely submerged. Anyhow, he had never been so comfortable in his life. Every inch of his body was supported. The finest innerspring mattress could not support him so evenly.

He was not swimming. His arms and legs lay motionless, resting. Why, anybody could do this — anybody, even if he had never swum a stroke in his life. All you had to do was lie there.

If you wanted to move, you didn't need to know

any professional swimming strokes. You could dog-paddle yourself along with your hands or, if you had fins on, any sort of kick would send you forward. Just to try it, he dog-paddled and kicked and moved smoothly through the water.

What a wonderful way for beginners to get used to the water! Fear is the greatest hoodoo of the beginning swimmer. He is so afraid of drowning that he can't keep his mind on his strokes. Using the snorkel, he wouldn't be afraid, and could work out his strokes slowly and carefully.

His paddling had taken him to a shallow part of the lagoon and the coral gardens lay only some ten feet below him. He floated above this lovely landscape as if he were in a helicopter, or on a magic carpet.

Below him the coral heads stood up like castles with holes that looked like doors and windows. Others were more like fine palaces. Fish wearing costumes as colourful as those of the knights and ladies of olden times passed in and out of the castles and palaces.

The battlements seemed to be moss-grown and ivy-covered. Roger knew that most of these graceful swaying things that looked like flowers and ferns were really animals.

No real castle walls ever had such a colourful covering. Many of the colours were new, colours seldom if ever seen in the land world, colours that you couldn't name.

Now his magic carpet took him over trees of a staghorn coral. At least they looked like trees. But he knew that their trunks and branches had been built by millions of busy little coral polyps. And there was a huge round thing that he took to be a brain coral. The folds on its surface looked just like the convolutions of the human brain.

He had seen these things in his brother's books on sea life. But for every one thing that he could name there were twenty that were a complete mystery to him. He made up his mind that he would learn all about them.

He did know the sea urchin and the sea porcupine and he was glad that he was floating above them rather than forced to walk among them where they lay thickly scattered on the lagoon floor. The sea urchin's dozens of long black spines and the sea porcupine's short white ones stood up like needles and if your foot or your hand touched them you would be in trouble for weeks. The spines would pierce the flesh and break off and would have to be dug out. They would leave poison behind them that would fester and hurt.

He passed close over a high minaret of coral with a gorgeous purple-and-gold flower on its peak. Surely this really was a flower. It had dozens of softly curling petals. He reached out his hand to take it and the petals all sank back and disappeared. He realized it was a sea anemone and the 'petals' were the tentacles that catch food and shovel it into the always hungry mouth.

He was dazzled by the giddy colours of the angelfish, butterflyfish, parrotfish and a dozen other kinds that he did not recognize – pink fish, blue fish, brown fish, and swarms of brilliant tiny yellow fish that swam fearlessly close to his mask and seemed just as curious about him as he was about them. One pressed its face up against the window of his mask to get a better look.

Roger saw a big fish swimming towards him. A chill ran along his spine. He could not see the thing plainly yet. It might be a shark or one of those giant barracudas.

Then he saw that the monster was only his brother.

Hal was wearing mask, snorkel and fins but carried one additional item that made Roger turn green with envy. It was a sea rifle. Dr Blake had shown it to them and had impressed upon them how expensive and valuable it was. This big-game underwater gun was powered by a CO_2 gas tank and would shoot sixty times with one loading. It had a pistol-grip rear handle and a forward handle of the machine-gun type. Its long tube discharged a barbed arrow connected with the gun by a fifteen-foot wire so that the fish could not swim off with the arrow so long as you held on to the gun.

Hal was cruising along slowly, watching for game. Presently he spotted a large grey snapper swimming among the coral branches. He took aim and fired.

It was a good shot. The arrow went straight through the fish and projected on the other side. The barbs would prevent it from being withdrawn.

The startled fish shot away but was brought up short by the fifteen-foot wire. Hal could feel the strong tug on the gun but he grimly held on.

The gun was whisked this way and that as the fish at the end of the wire shot up and down, in and

out, struggling to escape. The rifle bumped against Hal's mask and knocked it from his head. The mask sank out of sight below some coral branches.

Without the mask, Hal could not see clearly. He could not breathe either, since he had now been pulled under the surface by the fighting fish. To prevent the powerful fish from towing him away through the lagoon, he went to the bottom, got a grip on a coral head, and held on.

Roger was swimming over to help him. He thought he heard a muffled roar somewhere but did not stop to think that it might be an approaching motor-boat. His mind had no room for anything but this undersea drama.

He did not hear the shouts of Dr Blake who was hopping up and down like a madman on the deck of the *Lively Lady* and yelling at the top of his lungs. Shouts above water do not go under water. If just one of Roger's ears had been above the surface he would have heard the warning, but his head was submerged.

The brown men in the motorized fishing-boat stopped their laughing and singing to listen to the shouts of the man on the schooner's deck. But, knowing no English, they did not understand him. A split second too late one of them saw the end of Roger's snorkel projecting above the surface.

Hal, occupied though he was in his fight with the big snapper, noticed the black shadow sweeping towards Roger and heard the churn of the screw. Roger, swimming towards Hal, was coming straight into the path of the deadly hull.

Hal swam up towards his brother, but he could

make little headway because of the pull of the fish on his gun. It was a choice between saving the gun or saving Roger. He let go of the gun and the big grey snapper promptly swam away, towing the precious sea rifle behind him.

Hal crashed into Roger, violently pushing him out of the path of the oncoming boat. Then he ducked, but not in time to escape the boat's iron-ribbed keel. It struck him squarely on the head and then scraped over him as it sped on. His last thought before he faded out was that the blades of the propeller would chop him into mincemeat. Fortunately the men had already cut the motor and the quiet blades did no more than give him a good scrape.

Roger swam to his unconscious brother and held his head out of water. Dr Blake was already swimming out and the fishermen had plunged in to help. Blake and Roger, with the assistance of the natives got the inert body to the schooner and hoisted it on deck.

Blake felt Hal's pulse.

'Just knocked silly. He'll come out of it all right.'

He went below to get the medicine and bandages for the cuts and bruises. Roger and the natives turned Hal over a capstan and got some of the water out of him. Hal began to breathe in gasps. He opened his eyes to find Dr Blake looming over him. A look of complete disgust was on the doctor's face.

'Sorry,' Hal said, but Blake did not answer. He stooped and began patching up the battered body.

Hal felt as if he could sink through the deck for

shame. He had lost a costly gun, lost his mask, lost his fish, failed to watch out for surface boats. He and Roger were morons. They had been so anxious to prove their ability to the head of the expedition. They had botched their chance.

Hal expected that at any moment the scientist would explode and tell them what he thought of them. He almost wished he would. That would be better than keeping it bottled up and boiling around in his brain.

Blake looked daggers but said nothing. He barely spoke the rest of the day.

After they were all in their bunks that night, he said, 'Hal, you'd better get over to the airport tomorrow morning and meet the seven o'clock plane. Inkham will be coming in.'

'Inkham?'

'Didn't I tell you? I arranged for him before I left Honolulu. About your age – but he's really had experience in undersea work. I've seen him dive. He's good.'

Blake was silent a moment, then he added, 'It'll be good to have somebody around who knows which end his head is screwed on to.'

With which bitter reflection he turned over and went to sleep.

Hal lay awake all night.

2
The Practical Joke

Shortly after sunrise, Hal had the dinghy lowered. An outboard motor had been clamped to the stern of the little boat. Hal jumped in, gave the motor a twirl, and sped away across the lagoon.

It was a morning to make anyone happy. The sun shone gloriously. The water was as smooth and clear as a sheet of plate glass. The coral gardens at the bottom blazed with colour. The green islands towered a thousand feet high. Far away the surf broke white on the reef that encircled the great lagoon.

Anyone should be happy, but Hal was not. He still smarted from the humiliation of the day before. He had expected to be Dr Blake's right-hand man. But Blake thought he was a fool. Hal was inclined to agree with him. He had certainly made a brilliant donkey of himself. Now a new man was coming – a man Blake could trust.

Hal puzzled over the name – Inkham. Where had he heard that name before? It was an odd name, not one you hear every day. He searched his school memories, but in vain. All he could remember was that there had been something unpleasant connected with that name.

It was eleven miles to the airfield on the big island of Moen. The boat sped through a maze of small islands, passed large Tarik, Param and Fefan, then skirted the shore of Dublon littered with the ruins of the Japanese city destroyed by the bombing planes of the Allies during World War II. All these west Pacific islands had previously been ruled by the Japanese and were now a Trust Territory of the United Nations, administered by the United States. On Moen was a US Naval Base and airfield.

A plane droned in from the east and circled the airfield just as Hal pulled up to the dock and climbed out. He was on the field before the plane taxied to a stop.

Several men in naval uniform stepped out, then a young fellow in civvies.

Hal disliked his face the moment he saw it. And he was sure he had seen it before. That smart, sly, crafty look was hard to forget.

The newcomer stopped and looked around. Hal stepped up to him.

'Is your name Inkham?'

'S. K. Inkham, at your service.'

Then Hal remembered. 'Why, of course – I *thought* I knew you – you're Skink.' He thrust out his hand.

Skink took it, but without enthusiasm. 'And you're Hal Hunt,' he said sourly. He did not seem at all pleased to meet an old acquaintance.

To relieve the strain between them Hal said, 'Well, come along. I'll give you a hand with your bags. The boat is over here.'

As they walked across the field Hal's mind worked fast, digging up memories. He and Skink had gone to rival high schools. Skink's first name was Sylvester, but he didn't like it, so he always called himself S. K. Inkham. His fellow students couldn't swallow that, so they simply put together his two initials with the first three letters of his last name and nicknamed him Skink.

Hal could understand why Skink was not pleased to meet anyone who knew his school record. It had not been so good. He had been dropped from the football team for dirty work and suspended from his classes for cheating in exams. Then he had almost killed the biology teacher. The incident had created quite a sensation in the town.

The teacher had severely reprimanded Skink for the theft of a microscope. In revenge, Skink had dropped a sidewinder into the teacher's pocket. The adult sidewinder is only a foot long but it is a true rattlesnake and can inflict a deadly bite. The teacher put his hand in his pocket and was bitten. He spent three days in the hospital, close to death.

Skink was expelled from school. The Inkham

family moved to another town where their past was not known.

No wonder Skink was not tickled to bump into someone who knew him then.

Hal tried to make conversation. 'Well, how do you like our lagoon?' The boat spun a winding path between islands that looked like green towers spilling blossoms and fruit from all their balconies.

Skink looked around, and grunted.

Hal could guess what was on Skink's mind. He was afraid Hal would tell what he knew of him.

Should I tell? Hal wondered. The doctor has a right to know what sort of man he has on board. This fellow will cause trouble sooner or later. He might even wreck the expedition. I could prevent that by putting Dr Blake wise now. If Blake knows, he will drop Skink from his staff. At least he won't put him above me. I don't think I could stand having him lord it over me.

But he knew he wouldn't tell. Not even to Roger. Roger wouldn't remember Skink — he had been too young at the time.

Perhaps Skink had reformed. Perhaps he was really a good egg now. He should be given a chance to prove it.

'Look, Skink,' Hal said. 'I don't quite know how to say this — I think you and I ought to have an understanding.'

Skink looked at him suspiciously. 'What sort of understanding?'

'You had some bad luck in school. But you needn't think I'm going to blab about it.'

'I didn't get a fair deal in school.'

Hal thought about this. 'Seems to me you got more than a fair deal, Skink. You might have been tried for attempted murder. But your teacher refused to prefer charges. He even paid his own hospital bill. He insisted that what you had done was only a practical joke.'

'That's all it was,' maintained Skink. 'Just a joke.'

Hal could not answer. He could only look unbelievingly at this rascal who considered killing or near-killing only a joke. He thought of the days ahead – of the work under water. There were enough dangers down below without having this kind of joker in the game. But that was a chance that would have to be taken.

'All I want you to know,' he said, 'is that you're going to have a fair deal now.'

'Hunt,' exclaimed Skink, 'come down off your high horse. Who do you think you are, talking to me like a father? I can run my own affairs. Pretty soon I'll be running yours too. I know more about this undersea stuff than you and your Blake put together. Within a month I'm going to be boss of this expedition. Never mind my deal – start worrying about what kind of a deal you're going to have. If you were smart you'd get out now. If you don't, you can expect to take orders from me, and they won't always be nice. *Now* do we understand each other?'

'I think we do,' Hal said. His steady gaze bored into the shifty eyes of his companion. 'You want it to be you against me. All right, if that's the way you want it, that's the way it will be.'

They came alongside the *Lively Lady* and went aboard. Blake was at the rail.

'Good morning, Inkham,' Blake said heartily.

Skink was all smiles now. 'Nice to see you again, Blake.'

They shook hands. Dr Blake's eyes ran admiringly over the strong, supple figure of the new man.

'Good to have you aboard,' he said. 'We've been having a tough time – we need you.'

'I think I can help you out,' said Skink with a confident jerk of his head.

'Come below and I'll show you where you can put your things. Then we'll have breakfast.'

They went down the companionway into the cabin. There was a smell of hot coffee. Omo, who doubled as seaman and cook, was putting breakfast on the table. Blake moved toward the rear of the cabin.

'You'll sleep here,' he said, indicating a bunk in the stern. The bunks were close together there and the headroom was not so good.

But Skink had stopped in the roomier part of the cabin beside the widest bunk.

'Is this one occupied?' he asked.

'Yes,' Blake said, 'that happens to be mine.'

Again he turned aft but Skink did not follow. 'I'd be mighty sorry to inconvenience you,' Skink said, 'but the fact is I wouldn't be much use to you if I slept back there. The motion, you know. Rolling I don't mind, but pitching knocks me out. I'd do better amidships. But of course I wouldn't think of disturbing you. I'll just sleep up on deck.'

'Not on your life,' said Blake generously. 'Take my bunk and I'll move back.'

'Sure you don't mind?'

'Not a bit.'

Skink threw his luggage into the master bunk.

'Now for something to eat,' said Blake. 'Of course we usually have breakfast earlier than this, but we waited for you. Here's Captain Ike now. Captain Flint, meet Mr Inkham.' They shook hands. 'And this is Roger.'

Skink said, 'Hi,' in the tone of one who has no time to bother with kids.

'Omo, Mr Inkham.'

The handsome young Polynesian came forward with outstretched hand, a smile parting his lips to show flashing white teeth set in a face as brown as mahogany.

Skink suddenly took an interest in something else and appeared not to notice the hand. Omo withdrew it and calmly went back to his duties. He showed no resentment.

But Hal was boiling with rage. His fist hardened, his muscles tensed, and he could hardly restrain himself from landing a smashing blow in the middle of Skink's smug face.

So, Skink thought himself too good to shake hands with Omo! Omo, who was worth a dozen Skinks. Omo, who had more than once risked his life for Hal and Roger. Omo, who had shown such patience and courage during those terrible days on the desert island and on the raft. The brown giant's education was probably equal to Skink's and he had something more important that Skink lacked –

character. Hal and Omo had sworn to be blood brothers, following the old Polynesian custom. And now that his 'brother' had been insulted, all he could do was sit and fume.

Never mind. There would come a day when Skink would answer for this.

The breakfast consisted of tropical fruits, turtles' eggs, toast and coffee. When it was finished Skink said,

'Now, Blake, you might brief me on your set-up. We didn't have much of a chance to talk in Honolulu.'

'That's right,' Blake said. 'You really don't know much about us, nor I about you. But I saw you dive, and that was enough. Anybody who can dive like that . . .'

'Thanks,' smiled Skink.

'You already know that I'm with the Oceanographic Institute to study sea life and collect specimens. But perhaps you would like to know about this schooner. She's a fine little ship, sixty feet over all, and carries the triangular Marconi sail, the fastest sail in the world. She also takes a jib and two staysails. She has an auxiliary engine to get her through tight passages. She is equipped with tanks for specimens.'

'How did she happen to be fitted up with specimen tanks?'

'Before I chartered her,' explained Blake, 'she had been used by Hal Hunt and his brother to take specimens for their father who is an animal collector. She belongs to Captain Flint here. When they finished their expedition, I chartered her from Flint on condition that he would come along to run her. And since these boys had sailed her, I employed them too.'

'So the Oceanographic has given you the power to hire and fire as you please?'

'That's right,' said Blake.

Skink smiled at Hal. It looked to everyone else like a friendly smile, but Hal knew what it meant. Skink intended to see to it that Hal and his brother were fired. Then there would be no one to tell tales.

'Besides collecting specimens,' went on Dr Blake, 'we're supposed to keep an eye out for sunken ships.'

Roger snapped to attention. Here was something to catch a boy's fancy.

'Treasure ships?' he exclaimed.

'Well, yes, you might call them treasure ships, although the main thing the oceanographers and historians want is not treasure, but information about how men lived and sailed in the old Spanish days. You see, from the sixteenth century to the nineteenth, all these islands were owned by Spain. So were the Philippines. Spanish ships, loaded with the gold of the Philippines, used to come by here, stopping for food and water at these islands, and sailing on to the coast of Mexico, which was also Spanish. There the cargo would be transported overland, then reshipped to Spain. Because the ships could touch at Spanish territory all along the way, it was safer than taking the other route around the world.

'But these old galleons were none too seaworthy, and many of them went down — along with all the interesting things they carried in their cargo. Some people think that stories of sunken treasure are just stories, but the truth is that thousands of ships lie at the bottom of the sea, waiting to be found. A large proportion of the Spanish losses were along this route because it lies in the path of the typhoons. Few of them have been located because diving technique wasn't good enough. But now, with all the new diving inventions, the aqualung, undersea sled, bathyscope, and the rest, we ought to be able to do a lot better.'

They went on deck. It would not do to dive too soon after eating. So they stood by the rail and looked down to the colourful hills and valleys of the coral landscape, indistinct now because of the depth.

'It's another world,' Blake said. 'Nothing like it in the top world. I've been diving for twenty years. Sometimes I think I feel more at home down there than up above. It grows on you. At first it seems strange and perhaps a little terrifying. There are dangers, of course, but there are dangers in crossing a city street. After nearly getting knocked down by flying taxicabs, it's a relief to sink into a world of quiet and peace. Have you ever read Jules Verne's *Twenty Thousand Leagues Under the Sea*?'

The boys nodded. They had all read it.

'Then you remember that when one of the crew of the *Nautilus* died, they buried him at the bottom of the sea. I've often thought of that. It's just what I would want.' Skink laughed a little but Blake went on, 'I'm quite serious. I have no wife or children, nothing to draw me back to the land. If anything should happen to me, I could ask for nothing better than to be put away in a quiet coral garden like that one.'

He laughed as he noticed their sober faces.

'Don't worry,' he said. 'I'm not moving down for quite a while yet. Now, let's get out the gear and plan the day's work.'

3
The Scorpion in the Helmet

It was decided that Hal should go down in a diving-suit. Dr Blake considered the diving-suit old-fashioned, but there were times when it must be used. The diving-suit was an old story to Skink, and Roger was thought too young to risk its dangers.

Hal admitted that he had never been in a suit and could do with a little practice.

Dr Blake ordered Captain Ike to move the schooner to a deeper part of the lagoon.

While this was being done, a heavy rubber diving-suit, heavy copper helmet, and still heavier leaden boots were brought on deck. Then came a great coil of lifeline and a still bulkier coil of air-hose. Then a pump, and a compressor.

A scorpion that had been hiding among the gear skittered away across the deck, its slender tail and venomous sting arched above its greenish-white body.

'Those things come on board in the baskets of fruit,' Blake said.

Hal drew on the clumsy suit. Within this waterproof, airproof garment he at once began to sweat profusely.

The suit was so bulky and heavy that he was unable to bend over to put on his boots. Blake fitted them to his feet. Each boot had a thick sole of solid lead and weighed fifty pounds. When Hal tried to walk he found he could scarcely lift his feet.

'Next, the helmet,' Blake said. 'But there's one valve missing. I'll get it.'

He went down the companionway to the storeroom. Roger was at the bow, attracted by the flashing of a porpoise. Hal was busy inspecting his suit. So there was no one to notice when Skink went to the scuppers where the scorpion had lodged, picked it up deftly by the tail, and dropped it inside the copper helmet.

Blake returned, and with Skink's help lifted the weighty helmet, let it down over Hal's head, and locked it to the suit.

The pump was started and air began to come through the hose into the helmet. Hal peered out between iron bars through the narrow window and felt like a prisoner in a death cell. The sun beating upon the suit and the metal helmet made him feel faint.

Was he going to collapse before he even got into the water? Then what would Blake think of him?

All told, the helmet, suit and boots weighed two hundred and fifty pounds. It was as if he were trying to carry a two-hundred-and-fifty-pound man. The perspiration rolled down his face. Leaning heavily upon Roger and Skink, he shambled over to the rail.

Dr Blake had lowered a short ladder. Hal sat on the rail and the three men helped him to get his heavy feet over and down on to a rung of the ladder. Then he slowly descended the ladder into the water.

The feet seemed to become lighter when they went under the surface. When his suit and helmet had also submerged, he was free of the terrific weight.

But he still felt like a prisoner awaiting execution. He could do little for himself. His fate was in the hands of the men above. If that pump stopped he was done for. If his hose buckled, he would get no air. If they let him down too fast he would get the 'squeeze', and if they drew him up too fast he would get the 'bends'.

And he could not forget that he had an enemy above, one who would stop at nothing to get him out of the way.

His feet touched bottom. He stood in a fairyland of grotesque coral figures, pink monsters, purple fans, blue and gold trees with branches like the antlers of a moose.

The air-line attached to his helmet and the lifeline fastened to his shoulder strap went up, up, and disappeared through the roof. It looked like a roof, a roof of frosted glass. He could not see through it. He could see the hull of the ship where it sank into the water, but above the waterline everything was invisible. He could not see Roger

peering down, or Dr Blake at the pump, or Skink playing out the lifeline and hose.

But he suddenly realized that too much was being paid out. As soon as he touched bottom the two lines should have been held taut. Instead, more had come down, the coils of the black hose and white lifeline lay on the lagoon floor beside him. He must be careful not to get tangled in those loops.

He practised walking. It was an awkward business. He had to lean far forward like a falling tree. It was a hard job to pull up a foot, advance it, and put it down again. The stiffness of his suit, now puffed up with air, made every move difficult.

Suddenly he received unexpected and unwanted help. One of the strong currents that stagger across the bottom of Truk lagoon caught him in the back and pushed him forward a dozen feet. He had no time to see what this did to his lifeline and hose. He had no sooner steadied himself than a reverse current carried him fifteen feet backwards and sideways.

He clung to a coral branch as the currents tried to make sport with him. With his free hand he pulled in the slack of his lines.

He noticed unhappily that the air-hose was tangled in a staghorn coral. Any pull on it against the coral cut off his air supply.

Then he felt something moving in the helmet. Something was crawling through his hair. It made a shiver run up and down his backbone.

He could not get at it with his hands. There was nothing to do but to keep on trying to free his air-line.

The many-footed thing was walking over his right ear now. He closed his right eye as it crossed his eyelid. It crept down his nose.

Now he could see it and what he saw made his blood freeze. It was a scorpion.

He had a wild desire to crash his head against the inside of his helmet to smash the evil creature. But he knew that at the least movement the scorpion would bury its sting in his face. The poison would flow into his flesh. It might not kill him, but might easily make him unconscious. Then he would fall, his air-hose would be buckled against the coral, and without air he would be done for.

Even if he smashed it with a quick blow it would have time in its death struggles to stab him. Suppose it plunged its lance into his eye? Then he would go around the rest of his life with one eye. If there was any rest of his life.

He must keep steady and cool. After all, he was used to dealing with wild things. He had let tarantulas and black widows walk over his hand. He knew that if you didn't bother a wild creature it wasn't likely to bother you.

So he tried to forget the thing that was now crawling over his lips and chin and concentrated on freeing his air-line. He edged forward in his clumsy armour to the staghorn around which his line was locked. He must do this job himself, for he knew he was too far down for those at the surface to see his trouble. He had been told to jerk on the lifeline if he wished to be hauled up. But he could not be hauled up until that air-line was loose.

Now the thing was on his throat and still going

down. If it tried to get through under his collarband it would almost certainly be squeezed and would strike.

Hal tried to control the trembling that made his hands unsteady as he worked over the tangled airline. It was maddening to feel the thing explore his collarband, go around his neck, and then around again. Every impress of its feet felt like the prick of a needle.

But now it was going up again . . . over the left jaw and cheek . . . over the left eye . . . across the forehead and back into the hair.

Hal was weak with relief. After you have had it on your face and eyes, it was nothing to have a scorpion in your hair.

It was moving about quite lazily now. Evidently it liked this jungle. Hal began to hope. With luck, he might get to the surface without being bitten.

But when they took off the helmet, would the scorpion get excited and go into its act?

Now he could feel it no longer. It was either lying quietly in his hair, or had gone on up into the helmet.

Hal felt he had been through thirty years in thirty minutes. 'Bet my hair's grey,' he chuckled to himself. He felt so nervous now that he could almost let himself go into hysterics.

Part of this was due to what is called 'rapture of the depths', a sort of intoxication like the intoxication of alcohol. It is caused by staying too deep too long while undergoing severe nervous tension. It is also called nitrogen narcosis since it is due to the effect of nitrogen on the central nervous system under pressure.

It makes men do strange things. They forget where they are. Their cares slip away and they begin to dream. They fancy that coral heads are mansions and the coloured fish are lovely ladies.

Everything was becoming unreal to Hal. He laughed and cried. He was very happy and nothing seemed to matter. He felt like lying down and going to sleep in the coral garden.

But some instinct kept his hands working on the air-line. Finally it came away clean. He reached for the lifeline and gave it a tug.

Then he passed out and floated away on billows of dreams.

When he came to he found himself lying on the deck of the *Lively Lady*. His helmet had been removed. They were working to pull off his boots and suit.

It was good to feel the sun after the chill of the water at the bottom. Good to breathe fresh air that couldn't be cut off by a kink in a hose. Good to feel the solid deck underneath.

Then he thought of the scorpion. His hand jerked up before he could stop it and his fingers ran through his hair. Nothing was there.

He began to laugh weakly.

'It's in the helmet,' he said. 'You'll find it in the helmet.'

'Find what?' Blake asked.

'The scorpion.' Again he laughed and tears stood in his eyes.

'He's silly with rapture,' Skink said.

Dr Blake turned the helmet upside down and looked inside. There was nothing to be seen.

'You'll feel better soon,' he said to Hal. 'You're imagining things.'

'I tell you there was a scorpion in that helmet. It walked all over my face. It nearly drove me crazy.'

Skink smiled at Dr Blake. 'They do think funny things when they get the rapture,' he said. 'Doesn't pay to send down men without experience. They cause more trouble than they're worth.'

Dr Blake nodded gravely.

'Another thing,' Hal said. 'Those lines. They weren't kept taut. A lot of slack came down. It got tangled around the coral. I had a devil of a time with it.'

A pained expression came over Dr Blake's face. 'Hunt,' he said, 'there's one thing we don't do on this ship. When we get a bad break we don't make excuses. We don't try to blame somebody else.'

The words shocked Hal out of his rapture. His mind cleared.

'I don't know what I was saying, but it must have been pretty bad. I didn't mean to make excuses.' He raised himself on one elbow. 'But if I ever find that somebody put that scorpion into my helmet I'll punch the living daylights out of him.'

'There was no scorpion in your helmet,' insisted Dr Blake. 'Roger, help me stow these things away.' They went below. Hal closed his eyes.

Skink picked up the helmet and looked inside. He seemed surprised to find nothing there.

In the wall of the helmet were several openings

that led back through passages to the air-hose. Skink went to the air-pump and gave it a few vigorous strokes, sending a strong blast of air through the passages of the helmet.

Now when he picked up the helmet, there was the scorpion. He dumped it into the sea. Then he replaced the helmet on the deck and went off whistling.

4
The Aqualung

Blake and Roger came up carrying the aqualungs.

'I believe you're familiar with the aqualung, Inkham,' Blake said.

'Oh yes,' replied Skink, with a characteristic toss of his head. 'I've done more than fifty hours with the aqualung.'

'Then you can teach Hal and Roger.'

Hal grimaced. There was nothing he would like less than to be bossed by Skink.

Inkham's chest swelled up like a pouter pigeon's. 'All right, fellows,' he said in a voice of command. 'Do as I do. First we put on our fins and our masks. Then the weighted belt. And now for the aqualung.'

He lifted the aqualung and threw it over his shoulder. The large cylinder of compressed air lay along his backbone. Just above the upper end of it at the nape of his neck was the regulator, shaped like an alarm clock. Attached to the regulator was a loop of air-hose just long enough to go around to the mouth and back to the regulator. A mouthpiece was set in the hose in front of the mouth.

Hal and Roger put on their aqualungs. Roger grunted a little, for the cylinder was heavy.

'You won't mind it when you go below,' Blake said. 'It weighs thirty-two pounds out of water, but only three pounds in water.'

'Now the mouthpiece,' commanded Skink. 'You fit the rubber flanges behind your lips and grip the rubber nubbins between your teeth. You'll find it's just like the mouthpiece of the snorkel. You breathe through the mouth just as you do with the snorkel. Now practise breathing.'

Roger's face purpled as he tried to get air.

'Inhale sharply,' Skink instructed. 'That will start it moving.'

Soon they were both breathing very comfortably. The air from the tank tasted exactly like fresh air except for the slightly rubbery flavour from the tube and mouthpiece.

'I hope you realize,' Dr Blake said, 'what a miracle you have on your back. I should say that, except perhaps for the submarine, the aqualung is the greatest invention in all the history of diving. We owe it to Captain Cousteau of the French Navy. Now for the first time since man's ancestors came out of the sea millions of years ago, man is able to go back into the sea and feel at home. With this thing, man is able to move about under the sea almost as easily as on land – more easily in some ways, because he is supported by the water. You have freedom. No heavy suit, no copper helmet and lead shoes, no lines running to the surface, no air-pump to get out of order . . . well, you'll see for yourselves.'

Hal removed his mouthpiece long enough to ask, 'How much air does the tank hold?'

Blake was about to answer, but Skink cut in. After all, he had been appointed teacher, and he wasn't going to let anybody else take his job, not even Dr Blake. 'The cylinder contains seventy cubic feet of air at two thousand pounds per square inch pressure,' he said, proud of his knowledge. 'It's good for about an hour below.'

'If you don't realize you've been down for an hour and the air suddenly gives out, what do you do?'

Hal addressed the question to Dr Blake, but Skink answered.

'Put your hand behind your back. You'll find a lever beside the cylinder. Press that and you get five minutes' more air – enough to get you to the surface.'

'Why should it take five minutes to come up?' asked Roger.

'Because,' said Skink, lifting his eyebrows to show that he thought the question very foolish, 'if you've been down deep, a hundred feet or so, you can't come up like a flash. If you do you'll get the bends. You have to stop two or three times and let your body adjust to the change in pressure. But you wouldn't understand about that.'

Roger glared at Skink. 'You're mighty smart, aren't you?'

Skink said sharply, 'Smart enough to teach you a thing or two.'

Roger was about to retort, but Dr Blake stopped

him. 'That's enough, Roger. No back talk. Now, all of you, get overboard.'

The boys climbed over the rail and down the ladder into the lagoon. Before their heads disappeared below the surface, Dr Blake called after them:

'If you see anything interesting down there, bring it up.'

Hal watched to see how Roger was getting along. He was about ten feet under. The bubbles rising from his aqualung showed that he was breathing regularly. Presently he swam off, as free as a fish.

Hal felt as if he were floating in air. In the joy of this new experience he forgot about Skink. The weight of the aqualung and the weights on his belt were just enough to keep him from rising or falling. He hung suspended.

He gave his fins a slight kick and was surprised to see how smoothly he slid forward. He did not need to use his hands.

He turned downward, and a few kicks sent him tobogganing towards the bottom. He turned up, and rose equally fast. When he stopped kicking, the momentum still carried him on a great distance.

He stood upright in the water, standing on nothing. He could stay here indefinitely, like a star in space. He noticed only that he rose a few inches when he inhaled, and sank a trifle when he exhaled.

This gave him an idea. He filled his lungs with a long deep breath. At once he began to rise gently through the water. Before he reached the surface, he emptied his lungs. Down he went, just as gently and surely. The discovery excited him. It was as if he had his own private elevator and could go up or down at will. He didn't need to move a muscle of his arms or legs. To rise or sink, he had only to breathe deeply in or out. His lungs were a balloon that carried him up or down, as he pleased. Normal breathing, which kept the amount of air in his lungs at about the same level, held him almost stationary.

He swam idly down and stood on the floor of the lagoon. He walked through the coral garden. His heart beat fast for the wonder of it. Think of it, being able to walk about freely on the bottom of the sea!

There was no need to worry about lines getting tangled in the staghorn coral, for there were no lines. He wasn't tied to something on the surface, like a dog on a leash. He was complete master of his own movements, he could go where he pleased.

How different this was from suit diving! Instead of a bulky, suffocating, rubber suit he wore nothing but bathing trunks. Instead of fifty-pound boots, he wore rubber fins like the wings of the Greek god Mercury. Instead of fretting lest someone up above might stop pumping air to him, he carried his air supply along with him.

He walked with a springy step that he had never known in the world above. The water lifted him along. The pull of gravitation had been reduced to almost nothing.

When he stumbled against a rock he did not fall but only swayed forward a little and then righted himself. He tried to fall, but found it was impossible. What an improvement this was on the upper world — this world where you could never fall!

He found that he had suddenly become a remarkable acrobat. His lightest step carried him several feet upward and forward. If he pushed the ground with his toes a little, he soared ten feet before his foot again touched bottom. He thought of the giant with the seven-league boots, and amused himself by taking long steps.

A coral head as big as a house rose before him. He gathered all his strength. He sprang and — wonder of wonders — sailed up twenty feet and more, over the head, and down to the bottom on the other side.

He had been good at the high jump in school. But he had never made a jump like this. He could never do better than five feet. The world record was only between six and seven feet. Jump that high on land and you're a world champion. And what tremendous energy it takes to do it! But the underwater athlete could leap four times as high with ease.

He came to the edge of a canyon that sank far down out of sight. The other edge of the canyon was thirty feet away. Hal leaped, soared across the terrible gulf, and came down as light as a feather on the other cliff.

It frightened him a little and he shivered as he looked back from the cliff's edge down into the black gorge.

But, he reminded himself, he had no reason to be afraid of a gorge. He emptied the air from his

lungs to make himself as heavy as possible. Then, mustering his courage, he stepped off the edge of the cliff.

Down he sank through water where red and yellow sun rays were replaced by cold blue and finally, black. His feet touched the bottom. It was very chilly here and his ears sang with pressure. He waited a moment for his eyes to become accustomed to the darkness.

Beside him he could see the face of the coral cliff pitted with deep holes. He saw several long, slender, waving objects. At first he took them to be tendrils of seaweed. Then he recognized the tentacles of a large octopus.

That was enough canyon for Hal. He inhaled sharply and struck out with both hands and feet to rise as rapidly as possible out of the horrible home of the sea devil.

He was still within the gorge when a black shadow moved over him. He looked up to see the outline of a great fish. It could be a shark — there were many sharks in the lagoon, attracted by the waste thrown from anchored ships.

The shadow remained poised above the canyon. Was it waiting for him to come up? It would wait a long while. Hal halted his ascent and tried to be patient.

He was getting cold. This wasn't fun any more. He hadn't reckoned on getting himself bottled up in a canyon with a shark above and an octopus below. Would the octopus come up to investigate? He looked down but could see no sign of the eight-armed monster.

But when he looked up again he saw that the shadow had moved closer. Now he could see plainly that it was a tiger shark.

He tried to tell himself that the shark wasn't really interested in him. But you could never tell what a shark would do. It might ignore you completely, or it might bite off an arm or a leg just for sport. He knew from his previous sea trips that even the so-called harmless sharks such as the sand shark and nurse shark sometimes forget their manners and inflict a deadly bite. And the tiger shark was one of the most fearless of the whole tribe.

Sticking to the shark's hide by their vacuum cups were two remoras or sucker fish. They were just going along for the ride. When the shark found something good to eat, scraps would float back to the remoras.

Out in front of the shark's nose was a quite different kind of fish. It wore vertical black-and-yellow stripes. It was a pilot fish.

It swam so close to the great jaws that the shark could easily have gobbled it up, but never did so. For the pilot fish was extremely useful to the shark. Its senses were very acute and it could often detect food when the shark could not. The little pilot would lead the shark to his dinner. Of course it was understood that the pilot would be paid for his services by a share in the meal.

Hal did not like the behaviour of the pilot fish. It would dart down in the mouth of the canyon, then back to the shark, then down again.

Finally the great shark languidly dipped and followed his small leader. This time the pilot did not pause at the brink of the canyon but swam down into it and the shark followed.

Hal knew it was time to move, but where? Desperately he cast about for a chance of escape.

Beside him in the face of the cliff were great holes, so common in coral reefs. He selected one that was about his size but too small for the big-headed shark. He swam in, his tank scraping on the ceiling.

Farther in the cavity widened and he was able to turn about and face towards the entrance.

Presently he saw the pilot fish leading the shark straight to the hole. The little pilot turned aside at the last moment, but the tiger came on. Hal shrank back as the great mouth, three feet wide, was pressed against the opening. Suppose the great shark were able to break down the coral and force its way in?

With the entrance completely blocked, the cave was as black as night. Hal lived through an agony of suspense. Gradually he began to hope that he had successfully baffled the shark.

But just as he was beginning to feel more at ease, he was startled by a blow on his leg. There was something else in the cave with him. It might be just a harmless fish of some sort. It might be something not so harmless.

Suddenly light poured into the cave. The shark had backed off, at least for a moment. But he still hovered only a few feet away.

5
The Giant Eel

Hal examined the walls and roof of his retreat. Then he discovered his companion. From a crevice near his left elbow two unspeakably evil eyes peered out. Below them was an open mouth set with rows of inch-long, incurved teeth like those of a boa constrictor. At the back of the mouth were gill openings.

But Hal knew that this was no fish. No fish in the seven seas could have such terrible eyes. Besides, the dark green skin was nothing like the scaly hide of a fish.

Hal knew that he was looking straight into the eyes of a giant moray eel. Being a good animal-collector, the first thing he thought of was not his own safety but the fact that this was one of the specimens Dr Blake most wanted to obtain. Of course, to be of any use in an aquarium, it must be captured alive.

He had no noose, no net, no narcotic drug. He had nothing but his two hands. And just outside the door waited a tiger shark.

But perhaps he could use the moray to get rid of the shark! The giant moray is the shark's most dreaded enemy. Even though the shark may be three times as large, it is at the mercy of the moray which can wheel and turn and twist so fast that the heavy fish cannot succeed in locking its jaws upon this wriggle of green lightning. The moray can take bite after bite from the soft underbelly until the shark bleeds profusely and other sharks come in to finish off what is left.

If he could just take the moray out with him he was quite sure that the shark would call it a day and disappear. He must grip the moray just behind the head exactly as he had often captured snakes. But he had never tried it with a giant moray. Those gill slits would be useful – if he could lock his fingers into them that would help him to hold on.

Suddenly both hands shot out towards the moray's neck. But the moray moved faster and the powerful jaws closed on Hal's left wrist. The sharp teeth bit painfully into his flesh. A thin trickle of Hal's blood oozed from the moray's mouth.

Attracted by the blood, the shark again pushed its great face into the cave entrance, shutting off the light. Hal tried to pull his arm away, but the teeth only sank deeper.

If he struggled he would lose his arm. He must

be patient. If this great eel followed the habits of other eels, sooner or later it would relax its grip in order to take a better hold. In that instant, he might wrench his arm away.

But it was agony to be patient under such circumstances. To make matters worse, the shark, excited by the smell of blood, began battering the entrance with his armour-plated head. Chunks of coral fell and the opening grew wider.

But suddenly there was a change of tactics. The big fish left the cave and swam away.

What Hal saw when he looked out made him wish that he could call the tiger back. For it was headed straight towards Roger and Skink who were just then floating above the canyon.

Skink turned and saw the oncoming fish. Instead of warning Roger, he left him to his fate, and put on all speed to reach the ladder. Then he swarmed up to the safety of the deck.

Roger, appearing bewildered, looked about him and discovered the shark which had stopped some twelve feet away. Hal was tortured with anxiety for Roger. Yet he held his arm quiet in the moray's mouth. He prayed that the creature would think the thing it held in its mouth was dead, and would loosen its jaws to get a better bite.

Then a new actor appeared on the scene. Dr Blake dived in, a shark knife held in his hand. It was a gallant thing to do, but Hal knew that the chances against Blake were a hundred to one.

Why didn't that kid brother of his swim for the ship? Blake was motioning him towards the ladder. But Roger would not desert the scientist as Skink

had deserted him. He drew his own knife from his belt and turned with Blake to face the tiger shark.

Unless Hal could do something, they were both quite sure to be killed. If they retreated now towards the ship, the shark would plunge after them. Their only chance was to frighten it by resolutely advancing towards it, and this they did. Sometimes that was enough to scare a shark.

But not this one. It stood its ground as they advanced. It opened its jaws in a sort of lazy yawn and the cavity was big enough to take in both of its enemies at one gulp.

Hal had been cold, but now he felt as if the sweat were streaming from every pore. It took tremendous nerve to keep his left arm still and limp. Presently he thought he felt a loosening of the jaws, but still he did not move his arm. He let it lie like a dead thing.

Suddenly the jaws opened and closed again all in a flash. But this time Hal moved faster, and the jaws harmlessly clacked together as both of his hands went around the moray's neck and his fingers gripped the gill openings.

At once pandemonium was let loose in the cave. The eel thrashed about wildly and its tail, which had thumped Hal during the blackout, now beat a tattoo upon Hal's legs. Too much of this would break a leg, for the moray could strike with the blow of a sledge hammer.

But its greatest desire now was to escape from the cave, and this happened to be Hal's most fervent wish also. By a common impulse, they shot out of the hole into the blue depths of the canyon.

Hal kept his grip on the creature's throat and locked his legs over the body, riding it as one would ride a horse. He twisted the green head upwards so that the eel was forced to swim towards the shark.

The big tiger was now slowly circling the two men with knives, watching for a chance to close in on them. Sharks are rather near-sighted, and it was not until the moray was within thirty yards that the tiger saw it. Then with a mighty slash of its tail it shot off through the lagoon, leaving two very much surprised and relieved swimmers behind it.

They were still more astonished when they saw what had frightened the shark. Straight towards them came a giant eel with a man from Mars on its back. It sped past them and collided with the ladder, Hal locking his leg around one of the rungs.

Blake and Roger hurried to help him. Dr Blake climbed to the deck and got a lasso, then went down again and fastened the noose over the moray's head. Hal still kept his grip on the throat as Blake and Roger, with the help of Omo and Captain Ike, hauled the struggling monster up on to the deck and into a specimen tank full of water.

It was noticed that during this time Skink stood at a safe distance.

The great eel looked like a furious sea serpent as it thrashed about in the tank, making the water boil. Dr Blake was delighted. 'It's nearly ten feet long,' he exclaimed. 'Wait till they see this at the Institute! Hunt, you're all right!' His hand fell on Hal's shoulder. Then he noticed the bleeding arm.

'Omo,' he called. 'Bring the first-aid kit.'

But Omo didn't need to be told. He was already coming up the companionway with a pitcher of hot water in one hand and an assortment of drugs and bandages in the other.

He helped Hal off with his diving gear, then bathed the arm. He put his mouth to the deeper cuts and sucked out the poison. Then he applied iodine, and a bandage.

'Dr Blake,' Hal said, 'thanks a lot for coming down just when you did.'

'Well, said Blake, 'when Inkham scrambled aboard with his eyes popping, I guessed you must be in trouble. By the way, where is Inkham?'

Inkham emerged from behind the mainmast.

Blake said contemptuously, 'It's safe for you to come out now, Inkham.'

'What do you mean by that?' Skink said sourly.

'I mean you have some explaining to do.'

'There's nothing to explain. Along comes a shark. I warn the kid but he's so scared stiff he can't move. I tried to drag him to the boat . . .'

'Did you see this?' Blake asked Hal.

'I saw it all. He's lying. He gave Roger no warning. He simply hot-footed it for the boat.'

'I thought so,' Blake said. 'You're a coward, Inkham.'

Skink bristled. An ugly leer came over his face.

'I don't take that from anyone,' he snarled. 'Stand up, Blake. Your time has come. I'm going to give you a lesson in manners.'

Blake rose to his feet. He advanced upon Skink. But Hal stopped him.

'Hold it,' Hal said. 'If you mess him up there won't be anything left for me to work on. After all, it was my brother he walked out on. Besides, there's another score to be settled. I'm guessing he put that scorpion in my helmet.'

Skink laughed loudly.

'You guessed right! And I wish it had killed you.'

Hal, who had been sitting on the deck, rose. When he was half up, Skink kicked him in the face and sent him rolling to the far rail.

It was just what Hal needed to tune every nerve and muscle for the fight. He sprang like a wild-cat up on to the boom and from this perch, he came down like a ton of lead on Skink's shoulders. Skink was flattened to the deck, but he squirmed like a snake and twisted himself on top of his enemy. Then he seized Hal by the hair and crashed his head repeatedly against an iron stanchion.

Hal, dizzy from the blows, still managed to get to his feet and land a hard blow to his opponent's midriff.

Skink doubled up like a jack-knife. Hal had a sudden idea. Before Skink could straighten himself out, Hal had leaped on to a plank that lay across the specimen tank containing the angry moray.

'Come on up,' he invited Skink. 'The eel gets the loser.'

Skink hesitated. His staring eyes went back and forth from Hal to the serpent-like monster that churned the water and leaped repeatedly, with razor jaws wide open, up towards the plank on which Hal stood.

Dr Blake laughed. This stung Skink into action.

He sprang upon the board and attacked with such fury that Hal almost lost his footing.

The two closed in a clinch, each trying to upset the other into the pool. The great eel beneath became more and more excited. Its wild leaps now brought its jaws closer and closer to the two strong twisting bodies.

The moray, like the octopus, is highly temperamental. In one mood it may be timid and retiring, but when thoroughly aroused it becomes a raging devil. What would happen to anyone who fell into that tank now, it was better not to think about.

Skink tripped Hal who fell across the board, his feet dangling on one side, his head on the other. He jerked his feet up as the moray made a pass at them. Then the moray shot to the other side of the board and leaped to bury its teeth in his face. It missed only by inches.

Skink deliberately put his foot on Hal's head to force it down within range of the poisonous teeth. Hal reached back and gripped Skink's ankle. A good wrench destroyed Skink's equilibrium and he fell with a terrified howl into the pool.

Hal realized what he had done. The infuriated eel might kill Skink. Even now it was preparing for a rush, its green head above the water, its evil eyes blazing.

As it plunged forward, Hal slid off the board into the tank. He caught the giant by the throat as it swept past, and was carried along by it. He struggled to turn the creature's head aside and give Skink a chance to escape.

He got sudden help from Dr Blake who appeared

with a landing net at the end of a long pole. He scooped the moray's head in the net. The powerful moray proceeded to tear it in pieces, but the delay provided enough time for Omo and Captain Ike to haul the screaming Skink out of the tank. Hal climbed to safety. Dr Blake withdrew the torn net.

Skink lay blubbering on the deck, still beside himself with fear and rage. Then, finding he was safe, his old arrogance returned. He stood up, dripping, shaking his fists at Hal.

'You'll pay for this,' he said thickly. 'Just wait. And you . . .' he addressed Dr Blake, 'you'll wish you had never seen me.'

'I wish that now,' said Blake.

'You think you're boss of this ship,' sneered Skink. 'You think you're going to order me about, make me dive for specimens, find sunken ships, bring up treasure – for you! Well, I'll do that, but I'll do it for myself. If there's any treasure coming up, it's going to be mine. I'm going to be chief of this outfit. And as for you, Blake, I see in my crystal ball that you're going to have a bad accident – a very bad accident.'

Blake laughed. 'Then it will have to be soon,' he said, 'because you're leaving on the next plane. I'm only sorry there's none for a week.'

'A week is plenty for what I have to do,' growled Skink, and lurched down to the cabin.

Blake shook his head. 'How mistaken I was about that fellow! It will be a pleasure to see him off on the plane.'

Captain Ike's weather-beaten old face showed deep concern. 'He threatened your life,' he said. 'If

I were you I'd fire him today. Then he can wait at the base till the plane leaves.'

'Nonsense,' Blake said easily. 'He doesn't mean half of what he says. Hunt scared him out of his wits, so now he's trying to make up for it with big talk. I'm not afraid of him. Besides, we need him.'

Captain Ike threw up his hands in despair. 'It's your funeral,' he grunted and went back to his work.

6
Wonders of the Lagoon

Dr Blake would not allow arguments to stop the work of the expedition. Soon the boys were again under water, and this time Dr Blake was with them.

Wearing aqualungs, they descended to another part of the lagoon floor. Here they stood in a forest of gigantic seaweeds towering fifty feet tall, and as slender as poplar trees. The currents made them sway as if in the wind.

Among the tree tops flew the birds and butterflies of this magical world — the butterfly fish with its spread of gorgeous wings much like those of its namesake in the world above; the flying gurnard and flying fish, both of which could soar through water or air with equal ease; the parrotfish with its green and gold colours and parrot-like jaws; the sailfish with its blue sail fluttering as if in a submarine breeze. An angel-fish hovered in the blue heaven. It did not wear the white that angels are supposed to prefer, but a splendid costume of yellow, blue, red and black.

There were even flying horses in this heaven. Sea horses perched in the branches by coiling their tails around the stems, or floated erect from one tree to another, fanning their tiny transparent wings.

The floor of the forest was no less beautiful. It seemed like a garden of rare flowers and plants. The plants were not merely green, as above, but glowed with every imaginable colour and many a colour not imaginable. Here were a hundred exquisite hues that had never been named. How could they be, since they were unknown in the land world?

The seeming plants were hard to the touch, for most of them were really coral. Here was crown coral, a crown that any king would be proud to wear — cup coral looking like a golden goblet — lacework coral that appeared to be as fragile as a cobweb but was made of tough stone — leather coral like an old saddle — organpipe coral that stood up in graceful columns. Tall sea whips waved back and forth. The explorers kept their hands off the thistle coral and the prickly alcyonia. Most especially they avoided the brilliant fire coral which could give you a rash as bad as a case of nettlerash.

Dr Blake stopped beside a large sea anemone. It looked like a big chrysanthemum, except that its dozens of pink tentacles did not remain still like the petals of a flower but kept constantly moving, reaching out for food.

If a fish or shrimp should rub up against those tentacles, something very strange would happen. In each tentacle were little threads like lassos that would be thrown out to grip the victim and paralyse it with poison. Then the tentacles would draw the dinner in to the waiting mouth.

But among the tentacles swam some tiny gold-and-black clown fish. They did not seem in the least afraid of the stinging lassos. They acted as if they were on completely friendly terms with the sea anemone. They passed close to the open mouth without harm.

Each of the hunters carried a net, without a pole, tucked in his belt. Dr Blake now removed his net, cast it over the anemone and pulled the prize loose from its coral rock. He swam upward with it, gesturing to the others to follow. Reaching the deck, he made the flower of the sea at home in a small tank.

Three little clown fish which had taken refuge in the mouth of the sea anemone presently emerged and swam about among the tentacles.

'If the anemone gets hungry,' Roger said, 'it can eat those little fish first.'

Dr Blake smiled. 'It will never do that, no matter how hungry it gets.'

Roger was incredulous. 'Why is that?' he asked.

'I'll show you. Bring me some angleworms.'

Roger went to the bait locker and brought a box of fat juicy worms. Dr Blake dropped one of them into the tank on the far side from the anemone.

At once one of the clown fish swam over to the worm and gripped it in its mouth. But instead of swallowing the morsel, the fish swam back and delivered it to the anemone. The tentacles clutched the worm, the poison stopped its violent squirmings, and it was passed into the anemone's mouth.

'But what does the clown get out of it?' Roger inquired.

'You'll see.'

Presently the clown fish that had brought the worm disappeared into the mouth of the anemone.

'He likes his food predigested. He'll go right on back into the stomach where the anemone's meal is being broken down by the stomach juices, and he will eat as much as he pleases.'

'But won't he be digested too?'

'No. He'll come out just as frisky as when he went in. There he comes now.' The little clown emerged, looking quite whole and happy.

'And watch the other clowns,' Blake advised. The two other fish were nibbling at the edges of the tentacles. 'They clean off the dirt and parasites. They keep their Aunt Anemone in good health. They even air-condition her. The fanning of their fins changes and purifies the sea water among the tentacles.'

'They help each other,' Hal said, 'like the pilot fish and the shark.'

'Right. And there are many other cases of the same sort of thing. The moray eel has a little companion that is allowed to enter its mouth. A rock fish has an attendant to pick off its parasites. Sometimes you will see a parrotfish balancing itself upright in the water while a certain small fish cleans its scales. And you probably know how the crocodile allows the crocodile bird to enter its open mouth and pick leeches and other parasites from its jaws. Nature is full of examples of teamwork. It's a pretty good lesson for men, isn't it?' He smiled at Skink.

But Skink wasn't accepting any show of friendliness. 'What is this, a sermon?' he grunted.

'It's whatever you care to make it, Inkham. Your inability to get on with others is going to make a mess of your life. I don't like to see it happen. But let's get back into the sea. I expect each man to bring up something interesting today.'

They sank again into the submarine forest and at once Hal saw something interesting. It was a fish that seemed to have no body. It appeared to consist of nothing but a large square head and two pop eyes. On the back of the head was a small tail.

It moved slowly and Hal was able to catch it with his hand. But now a remarkable change took place. The fish began swallowing water and with every gulp it increased in size. It was swelling up like a balloon. Hal had to use both hands to hold the growing ball.

Pins and needles seemed to be sticking into his hands. He saw that spines that before this had lain flat against the skin were now sticking out in all directions like the quills on a porcupine.

He remembered that he had seen one of these things in a museum. It was a hedgehog fish. He could not hold on any longer so he whipped out his net, put the pancake that had become a football into it, and took it to the surface.

Going down again, he saw another marvel of a quite different sort. At first he thought it was only a reflection or a shadow, for the thing was so transparent he could see straight through it. It was a big thing, six feet long, and it was crawling over the floor of the lagoon. He had no idea what it was. He was to learn later from Dr Blake that it was a sea lizard.

For a while it was only a pale white, like dirty window glass. Then it took on some pastel colours, yellow, green and pink.

It looked as if you could poke your finger straight through it, but when Hal tried he found it was quite solid. When he touched it the thing threw out long sticky threads that clung to his fingers. He took his hand away and tried to wipe it clean on the sand, but the gooey mess would not come off.

Hal's net was too small for this big fellow. He went up, got another one, and came down again. He saw that in the meantime the creature had swallowed some small fish. They were plainly visible, flapping about inside the stomach of the beast.

This was certainly a novelty that any zoo or laboratory would be glad to have. Hal trapped it easily in the big net. It had looked so light and airy that he was surprised to find how heavy it was as he towed it up to the surface and hoisted it on deck and into a tank with the help of his friend Omo.

Dr Blake came up dragging an enormous sponge. It was fully five feet long.

'I didn't know they ever came that big,' Hal said.

'Most kinds don't. But this is a very special sort, worthy of the king of the sea himself. It's named after him, Neptune's horn. I suppose that's because it's shaped like a gigantic trumpet.'

'Did you see my catch?' Hal pointed into the tank where the sea lizard lay.

Blake looked in. 'But this tank is empty.'

Hal laughed. 'Look again. Right down in that corner.'

Blake shaded his eyes against the sun and looked. 'Well I'll be . . . A sea lizard! Do you know that it's a very rare specimen? This is the first one I've ever seen alive. Congratulations, Hunt. You're a real animal man, and no mistake. Wish I had a dozen like you.'

In the meantime Roger was getting into trouble. Thanks to Skink.

They stood on the lagoon floor beside a group of corals. Roger was taking his instructions from the more experienced collector.

Skink pointed out the best corals and Roger put them into his net. A gorgonia was indicated, and plucked. Then a mushroom coral. Then a star coral.

Then Skink pointed out a red-and-grey object that looked much like the coral rocks around it. The boy reached for it, but some instinct made him draw back his hand at the last moment.

He looked more closely. The thing did not move and looked like a rough stone overgrown with bits of weed. It was about a foot long and there was a hole in one end of it.

Skink was gesturing to him to take it. But Roger was not as ignorant as Skink supposed. He had seen photographs of this thing and had listened open-mouthed to the natives of these islands who feared this creature more than anything else in the sea.

It was a stonefish, so called because it looked like a stone. It seemed harmless enough. But if Roger had put his hand on it the thirteen spines along the backbone would have pierced his flesh. Each was equipped with two poison glands.

The poison was as virulent as that of a cobra, as deadly as the venom on the tip of a head-hunter's arrow.

His flesh would have promptly turned dark blue. Within three hours his arm would be swollen to the shoulder. Within ten hours he would be delirious and running a high fever.

Victims of the stonefish suffered such agony that they tried to amputate their own limbs. They became insane, striking out at anyone who came near them. Many died within twelve hours, their muscles knotted and twisted with the pain and their faces so distorted that their friends could hardly recognize them.

The Polynesians called the creature The Waiting One. French settlers on the islands had a more terrible name for it, *La Mort* or The Death. Even the scientists could not be calm when they described it, and had named it *Horrida*.

Roger's first impulse was to leave the thing severely alone. But it would make a good specimen, if he could only get it up without being stung by it.

He took a broken piece of pipe coral and, using it as a stick, poked the stonefish out on to the open sand. Then he scooped it up in his net along with the corals. The imprisoned fish struggled to get out through the meshes of the net. Its poisonous spines protruded between the strands.

As Roger lifted the net, Skink lurched back in fear of being touched by it. He gave his fins a flip and disappeared among the sea trees.

But Roger did not go at once to the surface with his quarry, for he had noticed something else. It was a round, flat object with a very small tail. It lay on the bottom almost covered with sand.

'A sting ray,' thought Roger, and looked for the poisonous barb that the sting ray carries just where the body joins the tail.

There was no barb. Then this must be a harmless variety of ray. He took a second net from his belt. He would grab the ray by the tail and drop it into the net.

But when he touched the tail he got a violent shock. The thing must be an electric ray, or torpedo. This fish contains a battery that generates and stores electricity. The creature can turn the charge on and off at will. It delivers a jolt that can paralyse and kill a good-sized fish, but it is not fatal to man.

Roger had touched the tail only lightly, but it felt as if a dozen needles jumped into his hand. Now, though the stinging sensation had passed away, his hand and arm felt numb. Now he

understood another common name for the electric ray. It was sometimes called the numb-fish.

He wangled the ray into the second net.

He was about to go up with the two nets when a mischievous plan came into his mind. He thought of the devilish trick Skink had tried to play on him. The fellow deserved a good scare, and Roger believed he could deliver it.

He left the net containing the stonefish and coral to be brought up later. Holding the netted torpedo at a safe distance from his body, he swam in search of Skink.

Behind a giant coral toadstool he found him. Skink was bending over, rear end upward, as he explored a cavity in the coral.

Roger edged up behind him without being noticed. He swung the torpedo hard against Skink's thigh just below the edge of his bathing trunks.

Skink straightened up with a yell that blew out his mouthpiece. He clapped his hand to his thigh and looked around. He saw Roger and the net, and supposing that the net contained the deadly stonefish, he struck out frantically for the surface. Roger followed and managed to land another whack of the torpedo on the flying legs.

Hal and Dr Blake on deck were startled to hear wild cries of 'Help! Help! I've been murdered!' and jumped to the rail to see a half-crazy Skink, clinging to the ladder, gargling sea water, burbling and babbling something about having been stung by a stonefish.

They hauled him up. Screaming and twisting, he dropped in a heap on the deck. Roger climbed aboard, keeping his net and its contents out of sight.

'Quick!' screamed Skink. 'Take me to the hospital. I'm dying! The kid. He rammed me with a stonefish.'

He clutched his thigh. 'I'll go crazy with the pain. I'm going crazy now!'

Dr Blake pulled aside the clutching hand. 'Let me have a look.' He examined the area carefully. 'There's no sign of any puncture. And the flesh isn't blue. Aren't you mistaken in your diagnosis?'

'You want me to die!' bawled Skink. 'I tell you, get me to the hospital. Oh, oh, the pain! I can't stand it.' He was blubbering like a baby.

'Calm down,' said Dr Blake. 'Think it over – are you really in pain? Or are you imagining something?'

'The kid tried to kill me. I helped him get a stonefish. Then he struck me with it. I have just a little while to live. I'm getting delirious now.' He began to grovel about the deck.

Dr Blake seized him by the shoulders and pulled him up into a sitting position. He shook him smartly. 'Snap out of it, Inkham! Now tell me – do you really feel anything?'

Skink wore a puzzled expresion. He put his hand behind him and felt himself. 'Well,' he said defensively, 'I did when it struck. It felt like a thousand needles. But,' he looked more bewildered, 'I guess I didn't really feel it after that.' A new look of horror came into his face. 'But do you know what that means? I'm paralysed. That's why I have no feeling.' He tried to move his leg. 'See? It's numb from the hip down. I can't feel a thing.'

'Not even this?' and Dr Blake gave the leg a good pinch.

'Never felt it.'

It was Dr Blake's turn to be worried. He looked at Roger, who was holding his net behind him.

'What do you know about this, Roger?'

'He's right, there was a stonefish . . .' Roger began.

'You see?' yelled Skink. 'Now will you get me to that hospital? Or do you want me to die here?'

'He tried to get me to grab it,' Roger went on. 'I got it in my net. Then I caught this in my other net.' He brought the torpedo into view. 'I smacked him with it and he thought he was getting a taste of the stonefish. He was so scared, he swam up feet of a big shark and never noticed it.'

Skink staggered to his feet and advanced upon Roger. 'So you had your fun, did you? Now I'm going to have some fun. I'm going to give myself the pleasure of tearing you apart.' But his stiff leg refused to work and he pitched forward on the deck. 'I'm paralysed,' he whined.

'That numbness will pass off in a few minutes,' said Dr Blake. 'And don't take it out on Roger. You had it coming. In fact, I don't think you got half what you deserved.' He took the net from Roger and held the torpedo up for inspection. 'It's a dandy. Here's a tank that will suit it.'

Roger slid into the water and presently reappeared with the other net. Dr Blake was greatly pleased with the stonefish. 'There are many varieties and this is one of the rarest,' he said.

'The shark is still hanging around,' Roger said. 'There he is.' Fifty feet out from the ship two fins cut the surface. Beneath them could be seen the slate-blue back of the fish.

'Looks like a mako shark,' Blake said. 'It probably won't bother us if we don't bother it. I don't want the shark. But there's something I would like to have — that turtle. It's a hawksbill and a beauty.'

The turtle was swimming lazily on the surface off the starboard bow.

Hal was preparing to jump in. 'No use swimming after it,' Blake said. 'It can go faster than we can. It can outrace most fish, when it really wants to.'

'Could we catch it with the motor-boat?' Roger asked.

'No, it would just dive out of reach. I'm afraid we'll have to pass it up.'

Omo quit his job of splicing a halyard and came forward a little timidly. He was an excellent diver, but on this expedition he was supposed to act as crewman and cook while others took care of the diving.

'If you don't mind my trying,' he said, 'perhaps I could get the turtle for you. We have a way in the islands.'

'The field is all yours,' Blake said. 'Go ahead.'

'First I'll pay a visit to that shark.'

Without snorkel or aqualung, Omo slid soundlessly into the lagoon and swam down. They could see his brown figure pass under the shark. Suddenly the shark gave a startled thrust of his tail and swam away. Omo returned to the ship carrying something in his hand. He climbed on deck.

He held a remora. On the top of its head was the flat suction plate that it used to fasten itself to the hide of a shark. The remora will cling just as readily to certain other kinds of fish, or to a turtle.

Omo tied the end of a line through the gill and mouth of the fish. Then he went to the bow and spotted the turtle which was now about sixty feet off and getting farther away every minute. Omo took the free end of the line and made it fast to the rail. He hurled the fish far out so that it fell within a few yards of the turtle.

The remora lay motionless in the water, as if collecting its senses. Then it swam straight to the hawksbill and fastened itself to the big carapace.

Omo began to haul in on the line, gently, fearing to break the remora's hold. But it proved to be firmly glued to the shell. The turtle, sensing that something was wrong, suddenly put on speed. Its flippers beat the water in vain.

It tried diving. Omo let it go, but kept a drag on the line. As the turtle tired, he gradually drew it in.

A net was lowered until it lay a few feet below the surface. The turtle was manoeuvred over the net, and hoisted aboard.

Omo beamed, and everybody else cheered, except the sulky Skink.

'I learn something new every day!' exclaimed Dr

Blake. 'We think we're so smart with all our modern gear, but we can take lessons from island people who never saw a snorkel or aqualung.'

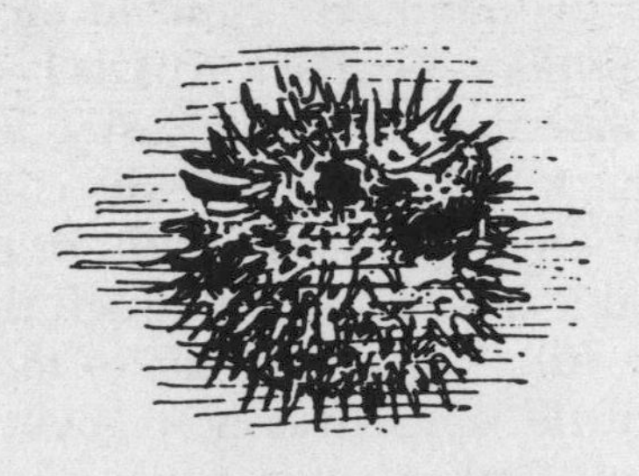

7
Are Sharks Dangerous?

The big mako shark had returned and lay just under the surface a few yards off the port beam.

'Wish he'd go away,' Blake said. 'It's a bit risky — doing any more diving while he's around.'

'He didn't make trouble when Omo grabbed one of his remoras,' Hal remarked.

'Omo took him by surprise. But he came back. The way he keeps switching that tail, I think he's a little annoyed with us. These mako can be man-eaters.'

'I heard a lecturer say that all sharks are cowards,' Hal said.

Blake laughed. 'Perhaps he felt safe because he had a good solid platform under his feet and there were no sharks on the stage. And even if sharks were cowards, don't forget that cowards are often bullies. Isn't that true among humans? I know I'm more afraid of a coward than of a brave man.'

Hal thought of Skink, and nodded. Yes, Skink was a good example. He was to be feared in spite of the fact that he was a coward. He was to be feared because he *was* a coward.

'But I wouldn't agree that all sharks are cowards,' went on Dr Blake. 'When a shark is hungry enough or mad enough it will attack a whale ten times its own size. It will even try to fight a ship. There have been numerous instances of sharks plunging their teeth into a ship's hull, sometimes even sinking the ship.'

'I suppose some sharks are more dangerous than others.'

'That's it. There are more kinds of sharks than of cats. And the man who says sharks aren't dangerous has probably met only the mild sorts. Besides, even the dangerous kinds aren't always dangerous. A shark that has just had a good dinner isn't interested in picking a quarrel. A tiger shark is as gentle as a kitten when well fed and a holy terror when starving. And sharks have emotions, just like people. If you go near them when they happen to be in a bad mood, look out.'

Dr Blake ran his finger along the outline of an ugly scar on his right foot.

'Another thing about sharks that makes them a lot like us,' he said, 'is that they make mistakes. I got that because a shark made a mistake. It saw my foot and thought it was a fish. Anything that flickers will attract a shark. That's why the Loyalty

Islanders tie a dark cloth over the soles of their feet when diving. The sole of the foot and the palm of the hand are generally brighter than the rest of the body. The shark can't see too well, and it may snap at the small flashing thing without realizing that it is taking on more than it intended.'

Omo, who was listening, said: 'I don't know why, but the place makes a difference. The sharks at Huahine never hurt anyone, but exactly the same kind of shark in the Tuamotus is a killer.'

'Perhaps they have plenty to eat in the one place, not much in the other,' Blake suggested. 'Or perhaps the Huahine people taught the sharks to be afraid of man, and the Tuamotu islanders didn't. Captain, what's your opinion? Are sharks dangerous?'

Captain Ike screwed up his wrinkled face and clamped his teeth on the stem of his pipe.

'I've known sharks for forty years,' he said, 'and the better I know 'em the less I like 'em. You can't make friends with a shark. Last time I was in Australia they gave me some figures: sixty-nine people killed by sharks on that coast in thirty years, one hundred and five wounded, two boats sunk, thirteen boats attacked.

'Fella down there caught one of those hammerheads that some people say are harmless. When he opened it up he found a human skull. Right over here at Ponape, next island to this one, they took a white shark. Its stomach contained a bag of money and the remains of a woman and child.

'And this mako . . .' Captain Ike looked over the rail at the sinister blue-grey form, 'he's a mean one.

He has teeth as big as shovels and sharp as razors. He's one of the fastest fish in the sea — and what a jumper! One of his favourite tricks is to jump fifteen or twenty feet into the air and come down wham! on a small boat and smash it to kindling.

'No,' he concluded, 'I don't trust sharks. Half of the time they'll run away from you. It's the other half of the time you have to worry about.'

The mako still waited. Lunch was called and everybody went below. When they came back on deck, the shark was still there.

Blake scowled. 'Perhaps he thinks this is his own special bailiwick. Well if he won't move, we will. Captain, let's try it over behind Tol Island.'

The captain up-anchored and, using the engine only, lazied the schooner eight miles down to the western part of the lagoon. There he dropped anchor in ten fathoms.

There was no sign of the shark. 'Believe we've shaken him off,' Blake said hopefully. 'The coral formations look interesting here. Let's see if we can get some pictures.'

The photographic equipment was brought up and Blake and Hal checked it with care. Hal was an ardent and experienced photographer, but this would be his first try at taking pictures under water. The cameras were a 35mm. loaded with colour film, a 2¼ × 2¼ reflex with black-and-white film, and a 16mm. motion picture camera. Each camera was housed in a watertight aluminium box with bronze fittings and a glass front.

Blake, finishing his work, went to the rail and looked about. He groaned. There, only twenty feet away, floated the mako monster. Its head was turned towards the ship and its beady eyes seemed to be fixed upon Blake. It was like a challenge.

Blake accepted the challenge. 'All right, old boy, people call you the man-eater. We'll come in and see if you live up to your name.'

He summoned his assistants for a conference. 'Since this big fellow won't go away, we'll use him. The Institute has been studying the habits of sharks and we can make a contribution by studying this one. We were discussing the question, Are sharks dangerous? Here's a good chance to find out. We can test the various methods of protection against a shark. Some divers put their faith in a knife. Others say a knife is no good – that a shark billy is better.'

'What's a shark billy?' Roger asked.

'A club – like a policeman's.'

'Would that have any effect on a shark?'

'It might – if you bang him on the nose with it. His nose is very sensitive. Some say you can scare a shark by shouting at it. Some believe air bubbles frighten a shark. Some think that it's just a matter of keeping your nerve – that the shark can tell when you're afraid. Then there's cupric acetate.'

'What's that?'

'It's a shark-repellent. Scientists learned that a shark won't touch a dead shark that has decayed. So they have taken some of the chemical that forms in decaying shark meat and combined it with a dark nigrosine dye to make small cakes sealed in waterproof envelopes. You attach one to your ankle. When you meet a shark you tear open the envelope and the cake dissolves. If it works as it's supposed to, the shark will turn up its nose at you and swim away.'

'I suppose,' Skink sneered, 'you plan to stay snug on deck while we go down and risk our lives making these fool experiments.'

'Don't worry,' Blake answered, 'I'll make the experiments myself. We must keep a record of the tests and the best possible record will be a motion picture. I don't order anybody to risk his life, but if there is someone who feels like volunteering to do the camera work . . .'

'That's for me,' interjected Hal, fearing that someone might get in a bid before him.

'Then what do I do?' complained Roger.

'I'd rather you stayed on board,' Blake said. 'This is not a game for boys.'

But Roger objected so bitterly to this arrangement that Blake relented. 'Very well, you can come in, but stay at a safe distance. Keep close to the ship. Have your knife handy and if we need you we'll signal. Inkham can stay there with you.'

Skink's jaw dropped. His eyes went to the waiting shark and his face paled. But he tried to put on a bold front.

'Nothing I'd like better than to take on that shark single-handed. But I'm afraid I'll have to miss the fun this time. My leg, you know — it's still so numb I wouldn't be able to swim. I'll have to stay on deck.'

Blake nodded. 'Sorry your leg is bothering you again. It seemed all right when you went down the companionway to lunch.'

'Yes,' admitted Skink, 'but you use a different set of muscles for swimming, and they're still paralysed.'

'Perhaps it's your nerve that is paralysed instead of your muscles,' suggested Blake.

Skink began to bluster but was interrupted by the appearance of Omo carrying a blazing acetylene torch. It had been adjusted for underwater work. Over the tip was fitted an air sheath through which compressed air would make a bubble extending out over the flame to protect it from the water.

'Where are you off to?' Hal asked.

'The captain wants me to do a little work on the keelson. The metal snapped when we struck one of those coral heads. It needs a bit of welding.'

He dropped over the side. The torch still blazed bravely under the surface. Omo disappeared under the hull.

Dr Blake, Hal and Roger put on their masks, fins, aqualungs and weighted belts. Each belt already carried a sheath knife and now a shark billy was thrust in beside it. Packets of cupric acetate were strapped to the ankles.

'But we'll make the other experiments first,' Blake advised. 'Don't open the envelopes until I give the signal.'

They descended the ladder into the lagoon. Blake swam slowly towards the shark, and Hal, armed with the motion picture camera, followed.

Roger unwillingly did as he had been told. He stayed near the ship. He did not enjoy being treated like a child. He was almost as strong as the other two, and as good a swimmer. Angry and rebellious, he almost hoped something would happen so that he would have to rush in to the rescue. He drew his knife and waited impatiently.

Dr Blake proceeded with his experiments. First he tested the theory that a shark will retreat if you show no fear and swim straight towards him. He advanced towards the mako. Hal started the camera.

The mako paid no attention to the approaching form until it came within ten feet. Then he moved his tail lazily and fell away to one side.

Again Blake advanced and again the mako moved out of his path — but not so far this time.

Upon the third advance, the mako did not budge. Blake stopped within five feet of the big muzzle.

The evidence seemed to be, at least so far as this shark was concerned, that it would retreat at first before a resolute advance, but that this technique could not be relied upon to scare the beast away.

Blake found himself uncomfortably close to the object of his study. But now would be a good time to test the bubble theory. He took a deep breath, then exhaled sharply and a great volume of bubbles rose from the regulator at the back of his neck.

Perhaps this might have frightened a smaller fish, but the mako was not disturbed. He seemed to be studying Blake as intently as Blake was studying him. Dr Blake began to feel like the specimen instead of the experimenter.

Blake began to move away. The shark immediately followed him. It kept the distance between them at about five feet. This was not enough for comfort, and Blake, becoming a little excited, struck out, splashing hands and feet.

At once the shark began to close in on him. It showed its instinct to attack anything that seemed to be wounded or afraid.

Blake bottled his fear and turned to face the shark, waving his arms menacingly.

At once the shark stopped, but now it was only four feet away.

Blake tested another theory. It was that a shark is more likely to attack at or near the surface because that is where it finds most of its food, helpless or dying fish, garbage from ships. At greater depths it is more wary.

Blake exhaled and sank slowly through the blue-green depths. The shark promptly came down after him but now did not venture so close. It began to circle him at a distance of fifteen or twenty feet.

Suddenly the shark noticed Hal who was still operating the camera from quite near the surface. The great tail gave one mighty thrash and the body shot up towards the big glass eye of the machine.

Mixed with Hal's fear was the thrill of photographing an oncoming shark. It loomed bigger and bigger and kept on coming. Hal kept his finger on the button and the film whirred through the camera. Now the great head filled the whole picture. Now a cave yawned as the monster opened its savage mouth, revealing rows of sharp white shovels.

With all his strength, Hal banged the heavy metal-encased camera against the brute's nose.

Promptly it changed course, sliding past him and scraping the skin from his shoulder with its sand-paper hide.

Hal turned to face another attack, but now he was joined by Blake who tested the merits of his shark billy by bringing it down with a resounding whack on the already bruised nose of the mako.

The shark swam away but immediately returned in a more deadly mood than ever.

Roger could not stay on the sidelines any longer. He swam in with his knife bared. He disregarded Hal's violent gestures warning him to stay out of range.

The shark saw him and came for him, its open mouth as big as a barrel. At the last moment, Roger twisted to one side and gripped the right pectoral fin. Hanging on to it, he was dragged along by the big fish. He sank his knife into the white underbelly. Red blood gushed forth.

Blake had clutched the other pectoral fin and his knife was sinking deep and often into the great carcase. Hal knew his duty as a photographer and kept the camera whirring. This was a picture of pictures.

The smell of the blood spreading through the water brought a sudden rush of new visitors. Sharks appeared from nowhere, from everywhere, ravenous beasts, fearless with blood lust.

Blake and Roger fell away from the bleeding mako and left it to the furious attack of its brother sharks. The pink water boiled with the thrashing of their great tails.

All would have been well if the savage creatures had kept their attention on the wounded mako, but in their fury they were ready to attack and devour anything. They lunged at the swimmers who wielded their shark billies and knives with deadly effect.

Blake ripped open the envelope on his ankle and signalled to the others to do the same. The cupric acetate spread a yellow tinge to join the pink of the blood-stained sea.

However much this repellent might have deterred a shark under normal circumstances, it had no effect whatever upon this bloodthirsty mob. The big fish were at too high a pitch of excitement to be discouraged by an unpleasant smell.

The three swimmers moved back cautiously towards the ship, fighting a rearguard action against the demented beasts. Here were mako sharks, blue sharks, white sharks and hammerheads, all of them intent on snapping up these human morsels floating in the pink sea.

Reaching the foot of the ladder, Blake seized Roger and made him go up first. But when nothing of Roger was left in the water except his feet, a mako lunged at those white fishlike things with such determination that Roger had to drop back into the sea to defend himself.

Above, leaning over the rail of the *Lively Lady*, was the laughing face of Skink. He was enjoying this spectacle enormously. Blake called to him to come down and help, but he blithely declined the invitation. No spectator ever enjoyed seeing the Christians thrown to the lions in a Roman arena more than Skink delighted in the death struggle of his three companions.

But he sang a different tune when a mako, making one of those high jumps for which the mako are famous, sprang a full fifteen feet into the air and came crashing down on the rail, smashing it to bits. The big body slid across the deck, scraping off generous portions of Skink's hide as it passed.

This was enough to remove any lingering numbness that might have remained in Skink's leg. He jumped like a jack-rabbit for the ratlines and swarmed up to the crow's nest. In this retreat he crouched, shivering lest one of the terrible acrobats of the sea might reach him even here.

Blake and Hal made another attempt to hoist Roger up the ladder, but again the sharks destroyed their plan. Roger dropped back into the sea.

The situation had become desperate. All three

swimmers had reached the limit of their strength and of their wits. The end could not be far off, and Hal found himself regretting that the wonderful film in the camera would sink to the bottom of the lagoon where no audience would ever view it.

Roger sank some distance to a point where, looking up, he happened to see Omo working on the far side of the hull with his acetylene torch, quite unaware of the battle being fought on the other side of the ship.

With powerful strokes Roger shot up to Omo's side and snatched the acetylene torch from the hands of the astonished crewman. Holding the flame-spitting machine, he swam under the keelson and came up into the churning mob of sharks.

Like King Arthur with the burning sword Excalibur, Roger attacked his enemies. The flame with its temperature of 3,600 degrees, a flame that could cut steel, was too much even for a blood-maddened shark. A big white shark limped away with a hole as big as a tub burned in the side of its head. In the time that it would take to open its mouth, a blue shark lost its lower jaw. The knight of the Round Table next took on a hammerhead which stumbled away with one of its hammers gone.

Here, there, up and down, flashed the deadly flame. The berserk fish came back to their senses, forgot about blood, forgot about everything except that scorching dagger, and fled for their lives in all directions.

Mute with astonishment, Blake and Hal waited at the foot of the ladder. There was not a shark in sight. Roger took the torch back to Omo, then joined them at the ladder. They climbed to the deck. The rail was smashed on both beams, where the leaping shark had landed and where it slid off again into the sea. From the crow's nest peered down the frightened face of Skink.

The three fighters dropped wearily to the deck. Hal set the camera down tenderly. In that camera was the greatest picture of a shark battle ever filmed.

Blake was looking at Roger as if he had never seen him before. 'My boy,' he said, 'I want to apologize for putting you on the sidelines. Why, you're a better man than any of us. Your wit saved us from a very messy death.'

Roger glowed under the chief's praise. He felt he had grown up. No longer would they call him a kid and push him off to one side when there was fun afoot. Now he belonged.

8
The Iron Man

'Today we'll try deep-sea diving,' Blake announced on the following morning. 'We want to get some colour pictures of life a quarter mile down.'

He smiled at the wide-eyed surprise caused by his words.

'You are aware, I hope,' said Skink scornfully, 'that the aqualung cannot be used at a depth of more than three or four hundred feet.'

'Quite aware. We won't use aqualungs. We'll use the Iron Man.'

Blake gave orders to Captain Ike and Omo, who removed the hatch and dropped a steel cable with a hook at the end from the tip of the cargo boom. Then the motor winch was started, the cable began to wind on to the drum, and up out of the hold rose a grotesque monster of steel and glass.

It had a huge head with four eyes, and a round body that reminded one of the belly of a very fat Santa Claus. The creature had no legs. But it had two steel arms, five feet long, and at the end of each arm were two steel fingers.

The monster was swung over and down to the deck. It seemed to be almost too much for the planking which sank a little under its weight.

'It weighs nearly two tons,' Blake said. 'The walls are solid steel, two inches thick.'

'Why do they have to be so thick?' Roger inquired.

'To withstand the tremendous pressure at great depths.'

Hal studied the monster with intense interest. 'Would you call it a diving bell?'

'That's right. But the very newest kind. The diving bell has a long history. Even the Greeks had a primitive one. But the machine had to wait until this century to become really efficient. You may have heard of William Beebe's descent in the bathysphere, and Otis Barton's benthoscope, and Professor Piccard's bathyscope.

'But the trouble with all these devices was that they were just observation chambers. You could get in and go down and look out through the windows, but that was all. If you saw something you wanted you couldn't reach out and pick it up. If you found a sunken wreck there was nothing you could do about it except observe it through the windows.

'Several attempts were made to fit diving bells

with arms and legs but they weren't too successful. A very clever robot invented by a man named Romano was used by Lieutenant Rieseberg in his search for sunken treasure. With its help he was able to bring up treasure from old wrecks. The machine you see before you is supposed to be the best of all these outfits, but we'll keep our fingers crossed until we've tried it.'

Hal was examining the steel fingers. They were long and sharp-pointed like the claws of a great bird. 'How do the arms operate?'

'By electricity. There is a switchboard inside for moving the arms in any direction and for working the claws. These claws operate like a pair of pincers. They can be brought together so delicately that they will pick up a small coin. Once you get used to them you can do wonders with them. I saw a demonstration in which an expert made the Iron Man's fingers tie a knot in a cable. And although they can do delicate jobs, the arms and fingers are very powerful. They can move great beams, or hatches, or trunks full of metal. They are at least twenty times as strong as the strongest human arms.'

Blake went around behind the monster and opened a heavy steel trapdoor, revealing a round hole about twenty inches across.

'Rather a tight fit, isn't it?' wondered Hal.

'Yes, but you can get through it if you slip one shoulder in before the other.'

They peered into the dim interior. In the head were the four round glass windows that, from the outside, looked like four eyes. The occupant would not be able to see up or down, but he could see out in four directions. There was room in this upper dome not only for a man's head but for a camera, if he wished to take pictures through the windows.

In the lower dome Dr Blake pointed out the switchboard by which the arms and fingers were controlled, other switches for spotlights to illuminate the dark ocean depths, cylinders that supplied air much on the principle of the aqualung, and the telephone by which the diver could keep constantly in touch with his friends on the ship above. There was even a small electric heater.

'A very necessary gadget,' commented Blake. 'It gets pretty cold away down there, beyond the reach of the sun's rays. Well, let's go out to deep water and make a trial dive.'

The *Lively Lady* sailed out of the lagoon through the western passage and on into the open ocean until the blue-black water beneath the hull told of great depths. There she hove to.

Dr Blake crawled into the Iron Man. The steel door was closed and bolted. The prisoner began to test the appliances. Hal, with earphones clamped over his head, heard Blake's voice: 'Is the telephone working all right?' Hal answered, 'I can hear you perfectly, Dr Blake.'

The spotlights flashed on and off. The arms began to move. Roger, who was within their reach, was suddenly caught between them and lifted like a feather from the deck, then put down again.

Then one arm wandered towards Skink and before that startled gentleman could shrink away the fingers snatched a handkerchief that had been tucked into his belt. The other arm descended to the deck and picked up a small nail.

Blake's voice over the phone was enthusiastic. 'It works like a charm. Put me over and lower away.'

Hal passed on the order to Captain Ike who started the winch. The Iron Man with the flesh-and-blood man inside rose some five feet from the deck, the cargo boom swung slowly out over the sea, and lowered the diving bell until it was just below the surface. The captain stopped the winch.

'Everything all right?' asked Hal. 'Any water leaking in?'

'Not a drop,' came the voice from the sea. 'Everything is shipshape. Lower away.'

The winch was started again and the Iron Man sank out of sight. A device attached to the drum measured in fathoms the amount of cable paid out. It ticked off ten fathoms, twenty fathoms, thirty fathoms.

Hal heard Blake's voice: 'She's riding very smoothly. The air pressure remains uniform. We have just passed through a school of mullet. They were very curious about this thing and stopped to look in the windows. One of them spiked himself on a finger but got away. It's getting dark now. How deep am I?'

'Fifty fathoms. Shall we stop?'

'Keep going to one hundred.'

At a hundred fathoms Captain Ike stopped the winch.

'You're at a hundred,' Hal said. 'What do you see down there?'

'Not a thing. It's as black as the inside of a

pocket. I'll turn on the spotlights. Ah, that's better! There are hundreds of fish around me — not the kinds we see in the top waters. The chamber is getting cold. I'm turning on the electric heater.' When Blake spoke again there was sudden urgency in his voice. 'You'd better pull me up. Water is leaking in around the door.'

'Hoist away!' Hal shouted. He was leaning on the rail looking down anxiously into the depths. Of course he could see nothing, but it seemed to bring him a little closer to the man below.

The winch had not moved. 'Hoist away!' Hal cried again and turned to see what the trouble was.

Skink was tinkering with the winch. Captain Ike had disappeared.

'The captain had to leave for a moment,' Skink said, 'so I took over.'

'All right,' stormed Hal, 'but pull him up. The bell is leaking.'

'Well now, that's just too bad,' drawled Skink. 'There may be a slight delay. This thing appears to be a bit out of order.'

'Well, fix it fast!'

'Don't you suppose I'm trying to?' whined Skink.

Hal still suspected nothing. He had too much faith in human nature to believe that Skink would be capable of plotting to drown Dr Blake. True, Skink had predicted that the doctor would have a very bad accident. But that was just a hollow threat, just big talk.

'What's the matter up there?' came Blake's voice over the wire.

'Something wrong with the winch,' Hal informed him.

'Call Skink. He's a good mechanic.'

'Skink is working on it now.'

'Tell him to hurry. The water is ten inches deep and coming in faster all the time.'

'Get a move on,' Hal called to Skink. 'Ten inches deep and gaining. The man will drown.'

'Well now,' Skink said easily, 'we wouldn't want that to happen, would we? Don't worry, I'll have this thing going again in ten minutes.'

'Ten minutes! You might as well say ten hours.'

Blake, who had evidently heard this remark, said, 'Ten minutes won't do me any good. This thing will be full in half that time.' His voice was quiet and matter-of-fact.

'He'll drown in five minutes,' Hal informed Skink. The latter was turned away so that Hal could not see his face, but he thought he heard a low chuckle.

Hal tore off his earphones and gave them to Roger. He drew his knife and leaped to stand behind Skink who was crouched over the winch. He touched Skink's bare back with the point of his knife.

'Don't move,' he warned, 'or I'll push this all the way home.'

'What the devil . . .'

'I tell you, don't move! Now, I'll give you just ten seconds to fix that winch. For every extra ten seconds this goes in half an inch.'

The winch began to run. Skink stood up and faced Hal. 'You didn't need to do that, you know,' he said reproachfully. 'By a happy coincidence I got it fixed just as you arrived. Don't flatter yourself that *you* had anything to do with it.'

Hal was embarrassed. He still could not believe that Skink had contemplated cold-blooded murder. Rather sheepishly he put away his knife.

The Iron Man rose above the surface and was landed on the deck. The door was unbolted and opened. A flood of water poured out.

Hal anxiously peered inside. 'Dr Blake, are you all right?'

'Fine as silk,' came a cheerful voice, and Dr Blake thrust out his head and one arm and shoulder. He didn't seem able to do more.

Willing hands seized him and pulled him out. He lay on the deck, pale, but smiling. He didn't waste any speech on the dramatics of the situation but thought only of its scientific aspect.

'Now that was very interesting,' he remarked, his voice shaking a little. 'At one hundred fathoms, six hundred feet, the water pressure was nineteen times what it was on the surface. That would kill a man instantly if he had no protection. But inside the diving bell I was just as comfortable a hundred fathoms down as I was on top — until the water began to come in. The more water that entered, the higher the pressure mounted inside the bell. It gradually made me numb — I suppose I have a mild case of the bends. But it just shows that if we can only keep the water out we ought to be able to go down a quarter mile without any trouble. We'll put some more packing around that door — then I'll try it again.'

'No you won't,' Hal said. 'Not today. You'll take a rest. It's my turn.'

Blake tried to get up but couldn't make it. 'Perhaps you're right,' he admitted. 'But anyhow you'd better get the water out of that thing. There's a valve at the bottom.'

The bell was emptied and dried and new packing was worked in around the door.

Hal took Captain Ike aside.

'While I'm down, I'd appreciate it if you'd stick to that winch. Don't trust it to anybody else.'

The captain understood. 'So you think there was foul play.'

'I wouldn't say that — I don't know. I only know I'd rather have you at that winch.'

'All right, if it will make you feel better. I won't let anybody else come within ten feet of it.'

'Good.'

Hal entered the chamber, taking the colour camera with him. As the bell sank below the surface he felt a momentary shiver of fear. But more than the fear was the thrill of entering a new and strange world, safe and snug in a steel cabin. For the next hour this was his home — a home under the sea — and why wouldn't it be possible some day for larger underwater homes than this to be built and for people actually to live in comfort in cities at the bottom of the ocean? Perhaps it was a fantastic idea, but many fantastic ideas had already come true. The land areas of the world were becoming crowded. Why shouldn't people move into the sea and make their homes there?

With proper protection against water pressure, it could be done.

The panorama from the windows was fascinating. A big ray lazily swam by, flapping its huge bat-like wings. Brilliant angel-fish flashed in the sunlight. One handsome fellow dressed in blazing colours came within four feet of the window. Hal took his picture.

A five-foot barracuda with all its dagger teeth showing circled the bell curiously. Hal was glad to be protected by two inches of steel plate. The barracuda suddenly made a rush and its jaws closed on a projecting bolt. Those teeth could cut through a wooden hull – but Hal had to laugh when he saw the barracuda's apparent surprise when those terrible teeth which could pierce anything that swam in the sea made no impression whatever upon this strange monster.

'Fifty fathoms now,' came Blake's voice over the phone. 'How about the leak?'

'Everything dry as a bone,' Hal reported.

The water changed from orange to blue, from blue to purple, from purple to black. The bell stopped.

'You're at a hundred fathoms. Are you still dry?'

Hal turned on the light and examined the rim of the door.

'Guess that packing did the trick. There's no sign of a leak.'

'Do you want to go on down?'

'May as well. This is just as comfortable as sitting on deck.' Hal turned on the heater.

A sudden underwater current struck the bell and

it began to turn. It went round and round. Hal didn't like it much. He began to feel strangely alone. Nothing connected him with the world of men but a half-inch steel cable and an electric wire. Where he was now, no one had ever been before since the world began. Perhaps he should not be here. He felt like an intruder surrounded by unknown enemies. The greatest enemy of all was the water pressure. How much would the Iron Man stand before it would collapse like an egg-shell? If that happened, death would be quick and painless.

Something much worse could happen. Suppose the cable snapped. The Iron Man would sink to the bottom and take up permanent residence there. Inside it, the soft man of flesh and nerves would enjoy no swift and painless end but would sit and wait in an agony of hope and fear until his air supply would run out and he would gasp his way into oblivion.

He idly wondered if the sealed chamber would embalm his body so that it would remain the same for hundreds of years, or whether there would be enough oxygen left in the steel tube to cause decomposition and leave only a skeleton at which some curious stranger would peer in a millennium from now when man should have made his home at the bottom of the sea.

He laughed off these grim thoughts, turned off the inside light, and looked out through the windows. The black sea was full of strange creatures carrying lanterns. Some hurried here and there, others like the softly glowing jellyfish waited for food to come to them.

There were white lamps, red, green, blue, yellow. It was like looking at night upon a city busy with traffic and the flashing of stop-and-go lights.

Some of the lights were sharp and clear, others were diffused and misty. Certain of the creatures Hal could recognize, having seen them brought up in deep-sea nets. The squids had lights around their eyes and on their tentacles. The shrimp shot out sudden streams of flame. The Venus' girdle carried a circle of light. One fish had illuminated whiskers. Another had no illumination except on two blazing rows of savage teeth which glowed because of a luminous scum that covered them. Deep-sea dragons carried rows of green and blue lights along their sides. The lantern fish had yellow headlights that he could turn on and off.

Hal told Blake what he was seeing. 'You might stop the bell for a moment,' he said. 'I'll try to take a picture.'

The bell stopped sinking, but it kept turning. The motion of both the bell and the fish made a time exposure impossible, and the creatures did not emit enough light for a snapshot. Hal used a fifth of a second wide open and hoped for the best.

'Wish this thing would stop revolving,' he said to Dr Blake.

'Sorry, there's nothing we can do about that. You're at two hundred fathoms now. Do you want to go deeper?'

Someone – was it the Iron Man? – told Hal to say, 'No, take me up.' But Hal did not say it.

'I don't know why not,' he answered. 'Everything is fine.'

Down sank the bell. Hal now turned on the spotlights and all the creatures that had spent their lives in eternal night were suddenly thrown into brilliant relief. Some of them showed fright and swam away. Others were curious and crowded up towards the lights. Hal took pictures until his thirty-six-exposure roll was used up.

Hal could hear excited voices on deck, then Blake said: 'You've made it! You're now a quarter mile under the sea – two hundred and twenty fathoms. Congratulations!'

'You can congratulate the Iron Man, not me. He's done all the work, and done it well. How about going a little deeper?'

'No, no, young man, you'd better let well enough alone. You're coming up now.'

There was a jerk on the cable, and the lights went out. Hal fumbled with the switches. They were quite dead. He did not hear the usual dull hum in the telephone. He called to Dr Blake. He got no answer.

Never in his life had he known such absolute silence. A quarter mile of insulation shut off all sounds, except those he made himself. His own breathing seemed noisy. He called again and was startled by the sound of his voice in the steel chamber.

He could guess what had happened – the electric wire had broken. The turning of the bell had twisted it until it snapped. Would the cable itself break next?

Or had it already broken? The bell might now be falling softly, silently, towards the ocean floor. That was believed to be three miles down at this point.

But no, when he looked out of the window he could tell by the lanterns of the fish that the bell was not falling. But it was not rising either. Why not? Had the engine failed again? Had Skink replaced Captain Ike at the winch?

Without the electric heater, the chamber was growing icy cold. It was plain that he would freeze to death before he would suffocate.

He called again, seized the telephone, shook it, tried to curb the panic that was rising in him. If he became excited he would only exhaust his air supply sooner. He must keep his nerve.

Suddenly there was a terrifying crash and he was thrown hard against the steel wall. There was a crunching, grinding sound as the bell scraped over the rocks of an underwater mountain peak. A current was swirling the bell about. Hal steadied himself and ran his fingers over the windows. They were not made of window glass but of the finest quartz and could stand great pressure but were not meant to take hard blows.

The bell was floating free again but the incident might be repeated at any moment. The ship above could not anchor in such deep water and had merely hove to. That meant it was slowly drifting with the wind, which Hal remembered was from the west. It was bringing the ship and the bell beneath it closer to the precipices that rose steeply from great depths to form the reef of Truk.

There were hatches that closed against the windows on the inside so that if they broke the sea could not enter. Hal tried to force them into position. They had not been greased and would not go far enough in to lock. They kept bobbing out.

He struggled with them for what seemed a long time but had to give it up. The effort made him warmer but as soon as he stopped to rest he began to freeze. It seemed to him that hours had passed since the wire broke.

Then he noticed that the windows glowed slightly like dim eyes. Perhaps it was just the phosphorescence of the fish outside. But no, it was different. It was daylight!

He looked out. The sea was changing from black to purple, from purple to blue, from blue to orange. The Iron Man broke the surface, rose into the air, and came down with a thump on the deck. A bolt screeched, the steel trapdoor opened.

'Are you all right?' It was Blake's anxious voice.

'Sure.'

Hands reached in to him. 'You're as cold as ice!' Roger and Blake hauled him out into the warm sunshine. He saw the electric wire twisted tightly around the cable and broken just above the bell.

'Did anything happen to the winch?'

'Things got rather balled up after the wire broke,' Blake said. 'Then we started hauling and you've been coming up at the rate of two hundred feet a minute. But you had a long way to come.'

He saw that Hal was shaking with the cold and the nervous strain of the terrible ordeal he had gone through.

'You must have had a rough time,' he said sympathetically, 'a quarter mile down, blacked out, no telling whether you'd ever come up.'

Hal tried to shrug it off – but his shoulders shivered rather than shrugged. 'I think I got some good pictures.' He lay down on the warm deck and fell sound asleep.

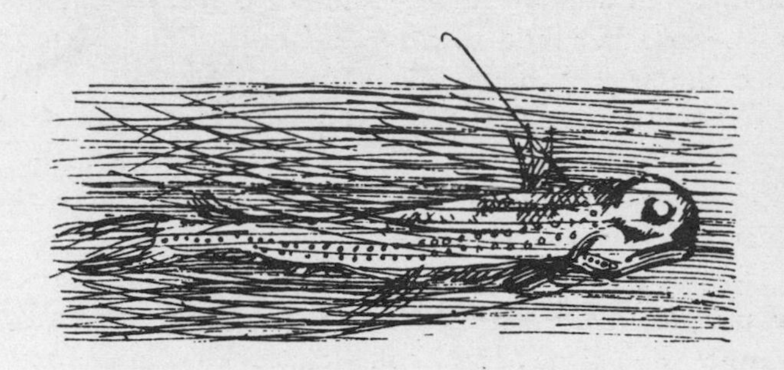

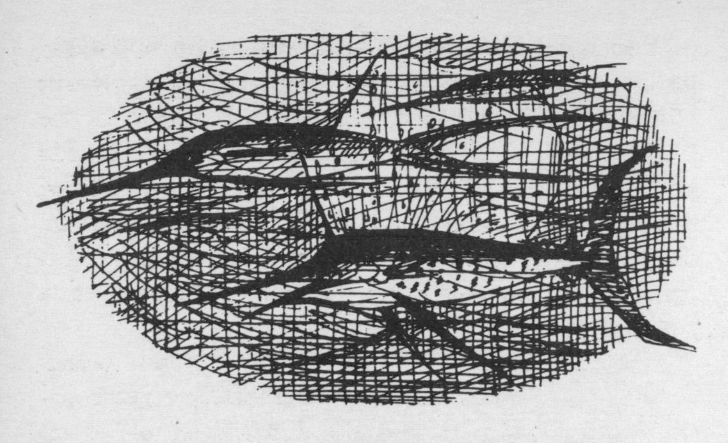

9
Treasure Hunt

It would be six days yet before Skink's plane would leave. During lunch Dr Blake announced his intention of sailing the ship to the island of Para for a few days, returning just in time to put Skink on the plane.

'Why are we going to Para?' Hal inquired.

'To search for treasure!'

Roger pricked up his ears. This was exciting news. He noticed that Skink was also interested, showing it not as the others did with delighted remarks, but with a dark scowl that made devilish lights flash in his cruel eyes.

Blake was not facing Skink and did not observe how he had taken the news. 'According to an old Spanish volume of travels,' he said, 'a galleon from the Philippines bound for Mexico and Spain foundered near the island of Para in the great storm of 1663. On board was the Spanish governor of the Philippines returning to Spain with all his household goods, gold and silver ornaments, tables, chests, statues, chandeliers, candlesticks, vases, bowls, cutlery – all the furnishings of a great mansion – cargo worth perhaps half a million dollars.'

Roger whistled. Skink's eyes glinted with greed.

'The Metropolitan Museum would like to obtain such articles to show how Spanish grandees lived three hundred years ago, and they have asked the Oceanographic Institute to be on the lookout for this wreck.'

'When would we leave?' Skink inquired.

'The island lies about a hundred and fifty miles south of Truk. We have a good west wind. The captain estimates that if we sail at sunset we should be there early tomorrow morning.'

'You still expect me to leave on the next plane?' Skink inquired blandly.

'Yes.'

'And you won't get back here until six days from now – just before the plane leaves?'

'That's correct.'

'Then I think I should go to the base this afternoon to book passage and make arrangements for my baggage.'

That seemed reasonable enough and Blake agreed. A look of sly satisfaction came over Skink's face and a sneer curled the corners of his mouth.

Only Roger saw it and it made him uncomfortable. 'What is the fox up to now?' he wondered.

The *Lively Lady* returned through the pass and crossed the lagoon to the eastern side of Moen. There she dropped anchor and Skink went ashore in the dinghy.

He was gone nearly two hours. The others put in their time watching the manoeuvres of half a dozen tiny submarines of the same type as those which invaded Pearl Harbour on December 7th, 1941. Known as one-man subs, but actually accommodating three, they had been built by the Japanese and left behind in Truk lagoon at the close of the war. Most of them had rusted away, but some had been refitted and improved by navy mechanics. One of the improvements was the addition of an escape chamber by which it was possible for a man to leave the submarine or return to it under the surface.

It was odd to see through the clear water a man emerge from a submarine, rise to the surface, go down again and re-enter, closing the trapdoor behind him.

'The escape chamber,' Hal explained to Roger, 'has two trapdoors, one opening into the submarine and one opening to the outside. If a man wishes to leave the sub, the chamber is filled with air from the sub's air supply. He enters the chamber and closes the door. Then the air is replaced by water to the same pressure as the sea outside. The man opens the sea door and comes out. He wears an aqualung, so he has no trouble breathing until he can get to the surface. Going back in, the process is reversed.'

'Wonder what's keeping Inkham?' Blake fretted. 'Booking shouldn't have taken him more than fifteen minutes.'

When Skink returned he seemed in high spirits. He did not apologize for keeping the ship waiting two hours, but stood by the rail enjoying the cavorting of the submarines while Captain Ike got the ship under way.

'I hate those things,' the captain growled. 'I can't forget what they did to us at Pearl Harbour.'

'I don't hate 'em,' Skink said happily. 'I love 'em.'

'They're no good for anything except mischief,' insisted the captain.

'That makes them very good indeed,' laughed Skink, and strolled off down the deck, leaving the old captain to bite the stem of his pipe and wonder what the devil the fellow had meant.

The *Lively Lady* flew like a bird all night and before sunrise dropped anchor in nine fathoms off the shore of the lovely atoll of Para.

It was a loop of land encircling a green lagoon only a half-mile long. The people of the island had fled during the war and it was now uninhabited. Its soil was volcanic rather than coralline and so rich that every sort of tropical tree and plant flourished magnificently — great coconut palms and sago palms, fantastic pandanus, stately bamboos, spreading mango and breadfruit trees, fruits and flowers of every description.

Somewhere in the waters round about this atoll

lay the wreck of the Spanish galleon, *Santa Cruz*. Dr Blake and his companions stood at the rail, looking down into the lovely blue-green depths.

'Are we the first to search for the galleon?' Roger asked.

'No. Many divers have tried to locate the wreck. Some of them died trying. Too bad they had to lose their lives – but every man has to check out some time and I can't imagine a more pleasant cemetery.'

Hal glanced up at the doctor's grave face. He remembered that the scientist had said something like this before. Evidently he was serious about it. His love of his mistress, the sea, to which he had devoted his life, must be very deep and profound.

'The reason that previous attempts have failed,' Blake went on, 'is that the divers could only go down and come up again. They could not stay down, move along the ocean floor, and inspect every inch of it. Now, with the aqualung, that can be done. But walking on the floor would be too slow. We must have a way to ride over it – and that's where the undersea sled comes in. Roger, do you and Omo want to bring it up?'

A strange contraption was hoisted out of the hold and laid on deck.

It was more like a surfboard than anything else, but it was narrow at the front end and spread out widely behind. Beneath it were two runners exactly like those on a sled. Roger found himself humming:

Jingle bells, jingle bells, jingle all the way,
Oh what fun it is to ride in a one-horse open sleigh.

But whoever wrote that song had never dreamed of going sleighing under the sea!

Dr Blake was explaining the mechanism. 'It's like a glider, except that it's built to fly under water instead of in the air. It was designed by an airman, Captain Vanlaer, an ace pilot in the French army during the last war. It's made of pressed wood and cork covered with a synthetic tissue. You notice it has twin rudders at the back, and also two ailerons like those of an aeroplane. With these controls the diver can regulate the depth of his glide. He can skim along the surface if he wishes, or descend to various levels, or slide along the ocean floor.

'The sled is towed by a motor-boat. Our dinghy with the outboard motor will do nicely. Even if we go along at a speed of only six knots we can complete the search of one square mile in half an hour. The same job done in the old way with divers going up and down would take the best part of a year. So you see this means a real revolution in sea-bed prospecting and the search for sunken wrecks.'

'Has it been used for that purpose?' Hal inquired.

'Not in the Pacific. In fact I think we will be the first to use it in this ocean. But it has been used for the past two years in the Mediterranean. First it was just thought of as a novelty, a toy for the playboys along the Riviera. Then its scientific value was realized and it has been the means of discovering eighteen sunken ships, some of them with valuable cargoes. They have found many aeroplanes shot down in the war. One of the men who tested it was Lord Louis Mountbatten – and the sled is now being studied by the British Admiralty with a view to using it in salvage operations.'

'I'm crazy to try it!' Roger burst forth.

'You'd be crazy if you did,' snarled Skink. 'It would be a good way to get yourself drowned. This is no job for amateurs.'

The remark irritated not only Roger but Dr Blake as well. 'I don't consider Roger an amateur. In fact, since he's the first to volunteer, I think we'll let him initiate the undersea sled.'

'Whoopee!' cried Roger, and leaped up to make preparations for the dive. Everyone helped to lower the submarine glider to the softly heaving surface of the sea. The dinghy was launched, a four-hundred-foot cable connecting it with the glider.

'It has to be long,' Dr Blake explained. 'Otherwise you wouldn't be able to go down far.'

Roger put on his mask and aqualung. He went down the ladder and, instructed by Blake, extended himself on his stomach along the glider. His feet slipped into the rudder controls and his hands lay on the levers that adjusted the ailerons.

'You'll find two straps fastened to the deck, one on each side of you. Put them around you and buckle yourself down.' Roger did so. Now he and the glider were one.

A button protruded from the sled's deck just in front of his face. 'What's the button for?'

'That's your signal. Press it.'

Roger pressed the button and a buzz sounded in the dinghy.

'If you want to stop, give me a buzz,' Blake directed, and climbed into the dinghy. Hal, a little anxious for his brother, stepped into the dinghy with Blake. The latter started the motor and idled the boat out four hundred feet until the slack in the cable was taken up.

'Are you ready?' he called back.

Roger took off his mask, spat into it, and rinsed it out. That would prevent fogging. He put the mask on again, testing it to make sure it was tight. He could guess that the rush of the water would do its best to tear the mask from his head. He adjusted the flanges of the aqualung mouthpiece behind his lips and closed his teeth on the rubber tabs.

He waved to Blake. The motor whined, the boat slid forward until the cable was taut, the sled began to move.

At first Roger was content to glide along the surface. Then he dipped the sled until the deck was awash and his arms and legs submerged, but his head still above. He dipped further and the water tore into his face. Involuntarily he blinked and held his breath — then realized it was not necessary to do either. The mask protected his eyes and although he was now completely under water he breathed comfortably from the tank on his back.

He steered down to a depth of about twenty feet. To stay down, he had to keep constant pressure on the controls. Whenever he let up, the sled immediately began to climb towards the surface. It behaved like the aerial glider, but in reverse. Whereas the aerial glider always tends to drop earthwards, the submarine glider wants to climb. Well, Roger reflected, this would be all to the good in case of accident. If the pilot should pass out, the sled would surface and be seen by the crew in the motor-boat. In fact, this was pretty soft compared with sky flying. It was a lot safer to fall up than to fall down.

He changed his mind a little about the softness of it when he was dragged through a great colony of jellyfish whose stinging tentacles left his skin on fire. But he would not signal to stop — this was too much fun. Besides, he longed to be the one to locate the wreck of the *Santa Cruz*. In a way it seemed fantastic to expect to find a wreck on his first dive. But why not? If the glider could cover in half an hour as much ground as could be surveyed in a year with old-fashioned diving methods, his chances were good.

The bottom slid swiftly by beneath him — not too swiftly, for the motor had been throttled down to six knots. He could see every detail of the ocean floor. It was covered by a thousand forms of life in all the colours of the rainbow. There were things like cabbages and roses, cauliflowers and lilies. There were fans and ferns and feathers. There were clouds of angel-fish, peacock fish and Moorish idols. He didn't like the look of the sea snakes in spite of their gorgeous livery of blue, gold and green on velvet brown. They glided in and out of holes in the coral or coiled around the branches.

Then there would suddenly be a stretch of snow-white sand, as bare as the desert. Then would come rock land, a wild tumble of crags and boulders.

He climbed sloping hills and went deep into the valleys to be sure to miss nothing at the bottom of the ravines.

He noticed especially the great number of giant clams, monsters four or five feet wide with their shells agape, waiting for food. If anything should come between them the shells would close like a steel trap. Many a diver whose foot had been caught in those merciless jaws had stayed to spend eternity at the bottom of the sea.

He shivered at the idea — but would have shivered more had he known the fate that one of these giants held in store for a member of the company of the *Lively Lady*.

At the end of about ten minutes Roger felt the sled swing around and then proceed in the opposite direction. Dr Blake had covered a mile and was now doubling back. He would keep doubling until Roger had surveyed a square mile of the sea bottom.

The floor here was flat and bare and looked like snow. Roger brought the sled down until the runners slid along the ground. Now he was actually sleighing at the bottom of the sea.

He went up over a slight rise, then tobogganed down a long slope. The slope ended in a precipice that dropped away to great depths.

Here was where a sleigh ride would end in disaster in the world above. Roger had a moment of panic as he shot out into space above the terrible abyss — but the sled soared like a bird over the yawning gulch and touched ground again on the other side. Roger's fear turned to exultation and he would have shouted with glee if he could have done so without losing his mouthpiece.

He was so happy that he failed to notice a hump

in the sand until it was too late. The sled ploughed into it and uprooted a large and very much surprised octopus. Having the ability to take on the colour of its surroundings, the beast was nearly as white as the sand. If it had been among brown rocks it would have been brown, or among green plants it would have turned green. But no matter what its surroundings may be an octopus always turns red when it is angry, and this one was a terrible red as it found itself caught on the prow of the glider and carried away at a speed of six knots.

Some of the tentacles were below the deck and some above, and two of them fastened themselves to Roger's bare back. The parrot-like beak of the creature was within a few inches of his face and the big, almost human eyes glared hatefully into his.

His first impulse was to signal for a stop. But if he stopped, the octopus could disentangle itself from the sled and attack. So long as he kept moving it might be too bewildered or terrified to do anything but hang on. The sac-like body of the beast was beneath the edge of the deck and the motion kept it there. Roger decided to keep going.

The two tentacles that lay on his body bothered him most. He felt them tighten, the suction cups biting into his flesh, in an effort to drag him forward within reach of the brute's jaws. The great beak opened revealing a hole big enough to take in a head larger than Roger's. Savage teeth edged the opening.

The octopus was disappointed, for the moment at least. It could not crawl towards him because of the water pressure that glued it to the moving sled, and it could not drag him closer because he was strapped to the deck. But what if the straps should break or slip?

Should he rise to the surface? Then the men in the boat would see him and stop. They would circle around and come to his rescue. But that would take several minutes, and in the meantime . . . Once able to move, the octopus would not need more than ten seconds to reach back and nip off his head.

No, he must stay down, keep moving, and fight this thing out all by himself.

The sled shot through a school of parrotfish. They were taken by surprise and several of them bumped into the octopus and into Roger's head and shoulders. He seized one of the big, fat, green-and-gold beauties and thrust it into the jaws that yawned in front of him. Perhaps if he treated his visitor to lunch it would lose interest in him. The fish disappeared into the black inside of the octopus.

The creature did not even trouble to close its jaws. Roger gave up the idea of winning his enemy over by feeding it. This was a case of anger, not hunger. The octopus, he knew, is a highly emotional creature. His companion on the flying sled was too mad with rage to worry about its stomach.

The sharp-rimmed suction cups of the two tentacles that lay along his back were tearing his skin. He felt himself pulled an inch closer to the waiting beak. He drew his knife and sawed away at one of the tentacles where it joined the body. It was as thick as a man's leg and as tough as rubber, but it had no bone in it. At last the great red snake was cut through, the suction cups relaxed their hold, and the tentacle fell away in the rush of water.

But another immediately took its place! Nor did the temper of the octopus seem to be improved by the operation. The body glowed a more fiery red than ever and the eyes burned with hate.

Roger felt the sled sweep round again and was reminded that he was looking for a sunken galleon. It was hard to keep your mind on it in the presence of such company. He laboriously sawed off another tentacle, and then another. Two more tentacles took their place. One of them strapped down his arm so that he could no longer use his knife.

He realized that he was breathing hard. That would not do. He would quickly use up his air supply, and then what would happen? He must breathe as lightly and shallowly as if he were sitting at ease on the deck of the *Lively Lady* instead of battling with an octopus on an undersea glider.

A dark shadow fell across him. He looked up to see that he was charging straight towards a cliff that towered fifty feet high and was covered with jutting crags and hooks of rock. He turned the sled upward. It responded sluggishly because of the weight on its bow. It rushed nearer and nearer the cliff, and the waving sea fans and great anemones attached to the base of the precipice loomed larger and every crevice and hole and out-reaching rock was visible.

If he crashed into it, that would be the end of the octopus. But it would also be the end of him and of the sled and of his hunt for the galleon. To protect himself he had to protect his unwelcome passenger. He steered upward sharply and just skimmed the summit, passing so close that the octopus was dragged through the plumes on the reef top.

Again he found himself puffing like a steam engine, and again he throttled down his fear and forced himself to breathe lightly. The two enemies stared at each other in silence for what seemed a very long time while the sled made another transit, and then another. Blood was drifting back from the animal's wounds, but it did not seem to be seriously disabled by the loss of three tentacles.

A new problem presented itself — seaweed in the form of great coils and festoons of giant kelp. Roger's wits were beginning to play tricks with him and he could imagine these long arms to be the tentacles of a monstrous octopus as big as a ship and with no purpose in life but to strangle and swallow Roger Hunt. He dodged here, there, up, down, to avoid the clutching fingers. He was weak now with fear and exhaustion. Then suddenly he was out of the kelp forest and sliding over a coral garden in which tall Neptune's sponges stood like Joshua trees.

Then he saw it — the wreck! Or, at least, it was *a* wreck. He could not be sure it was the *Santa Cruz*. It lay half buried in sand and covered with seaweed, barnacles and coral. He soared up over its broken-off mast and looked down on the high poop that certainly could not have belonged to any modern ship. His heart pounded with excitement. But there was time for only a glance and he was carried swiftly away. He dared not signal for a stop — not so long as he had this passenger. A dim shape loomed ahead. In another moment the sled would crash into it. Roger lifted the bow just in time to slide over the back of a big tiger shark.

Smelling the blood trail of the wounded octopus, the shark immediately turned and followed the sled.

It was soon joined by another inquisitive pirate, a great swordfish. Roger, glancing back fearfully over his shoulder, estimated that the sword alone of the monster was eight feet long.

He nervously expected the shark to come nipping at his white heels, laid out invitingly at the back of the glider. As for the swordfish, if it took a notion to, it could ram its sword clear through the sled, and through Roger as well.

He remembered the account of a swordfish that had rammed a schooner and driven its lance through a quarter inch of metal sheathing, three inches of Douglas fir, and two and a half inches of ceiling plank and had left its broken-off sword in the hull as a memento of this feat.

The swordfish came alongside on Roger's left and the shark drew up on his right. The three swam along together, like good friends. The octopus, no longer interested in Roger, twisted to the face of the swordfish, then turned again to fix its baleful eyes on the shark.

Even a shark is afraid of a swordfish, and with good reason. That mighty sword is one of the few weapons sharp enough and strong enough to pierce the shark's tough hide. The tiger kept at a safe distance and it was the swordfish that acted first.

It made a sudden rush, plunged its rapier all the way through the balloonlike body of the unfortunate octopus, and ripped it away from the sled. The octopus locked its five remaining tentacles around the body of the swordfish and there followed a titanic struggle that Roger could not wait to see. He was carried swiftly away, and was glad to go. A great feeling of relief poured through him.

But his nerves tightened again when he noticed that the shark, which had dropped back for a moment, had decided not to contend with the swordfish for the carcase of the octopus, but had turned its attention again to the sled. It was following close behind, probably admiring the white soles of Roger's feet and sniffing the scent that drifted back from the bow of the sled where the blood of the octopus had smeared the deck. In fact Roger himself was contributing a little to the blood smell from the suction disc cuts on his back.

The tiger could not be blamed for supposing that the strange creature fleeing from it was badly wounded, terrified, helpless, and could easily be turned into a good meal.

The sled swept around a curve and started back. Roger hoped the manoeuvre had shaken off his pursuer. But the shark was still close behind – even closer now.

What worried him almost as much as the shark was the chance of missing the wreck. On this transit he would not pass over it but should not be too far away to see it. He must get rid of this hanger-on so that he could give his full attention to his real job.

He thought of the way flying fish escape from sharks and other hungry monsters. They take to the air. Why couldn't he do the same? He had no idea how the sled would respond, but at least he could try.

The men in the boat were astounded to see the sled suddenly shoot up through the surface, rise into the air, soar for a moment, then plunge again into the sea. Before they could speak it happened again. And then again!

'That kid!' Hal exclaimed in disgust. 'He just must have his fun – doing stunts when he ought to be watching for the *Santa Cruz*! Sometimes I think he'll never be serious.'

But Roger was very serious. After the first two flights he could still see the shark some distance behind. After the third flight, he was alone. A moment later he caught sight of the wreck, some distance to his left. He signalled for a stop. He let the sled rise and skim the surface. The dinghy circled back and came alongside.

'What was all the jumping for?' Hal demanded angrily.

'I'll tell you about that later. I've found a wreck. Perhaps it's the *Santa Cruz*.'

Hal forgot his anger.

'Great! Where is it?'

'Right over there, thirty yards.'

'How deep?'

'About ten fathoms.'

The two men were about to dive in when Hal noticed the blood on his brother's back and on the deck of the sled.

'Why all the gore? Are you hurt?'

'It's nothing,' Roger said impatiently. 'Get in there and tell me if I've found anything.'

Wearing masks only, their aqualungs having been left on board, Blake and Hal plunged in, swam thirty yards in the direction Roger had indicated, and submerged. Roger unstrapped himself from the sled and climbed into the boat.

Within forty seconds the two were up again, spouting and blowing and red with excitement. They swam back to the boat where Roger waited in tingling suspense.

'You've got something there,' Blake said as he climbed in.

'Is it the *Santa Cruz*?'

'I couldn't make enough of an examination to be sure. We'll come back with the aqualungs.'

'How will we find it again?'

'That's easy.' Blake rummaged in a locker and brought out a line with a weight on one end and a flagged buoy on the other. He started the motor and eased the dinghy out to a point directly over the wreck. Then he dropped the weighted line. The buoy floated jauntily on the surface, waving its little red flag.

The dinghy returned to the ship. All on board were thrilled by the news, Skink as much as anyone, but in his own sour fashion. He could be seen scanning the horizon as if looking for an expected visitor. No one noticed him particularly, since all attention was focused upon Roger and his story of his undersea sleigh ride. Blake treated the cuts on his back.

'That was a fine piece of work,' Blake complimented him. 'You used your head. Now I suppose you want to know what you found.'

He went to the cabin and brought up a sheet of specifications of the *Santa Cruz*. He and Hal studied them carefully.

'All right, let's go and check,' Blake said. Taking their aqualungs, they set off in the dinghy, sternly refusing Roger's appeal to let him go along.

'You need to take it easy. We'll let you know soon enough.'

In half an hour they were on their way back. Roger, standing at the rail, could hardly wait for them to come within earshot before he called, 'How about it?'

Dr Blake stood up in the dinghy and cupped his hands around his mouth. His deep voice, softened by distance, came faintly across the water.

'It's the *Santa Cruz*!'

10
The Mystery of the Sunken Ship

'There's no doubt about it,' Blake said as he climbed on board. 'It's the ship we're looking for. Sunk three hundred years ago and still in beautiful condition!'

It was too good to believe. Roger said doubtfully, 'I should think a wooden ship sunk three hundred years ago would be rotted away by this time.'

'Not at all,' Blake said. 'You have to remember that the wood has been sealed away from the air all this time. If you bring some of it up into the air it will shrink and begin to decay rapidly. But so long as it is protected by the sea it will last not merely three hundred years but a couple of thousand. You know the book, *The Silent World*, by Captain Cousteau, the man who invented the aqualung. He tells of finding at the bottom of the Mediterranean, the Galley of Mahdia which sailed from Greece in 80 BC. The wooden decks and hull were still tight enough to hold together all the ship's art treasures which have since been transferred to the Museum Alaoui in Tunis. There were so many of them that they filled five rooms in that museum. Among them are some of the ship's rigs made of Lebanon cedar and covered with the original yellow varnish.'

'Does cedar stand salt water better than any other wood?' Hal inquired.

'Not necessarily. You probably saw in the papers a few months ago the story of the National Geographic expedition to bring up the treasures of a Greek ship that sank around 230 BC. They found the wood rubbery and tunnelled by shipworms but reasonably sound after nearly twenty-two hundred years at the bottom of the sea. The ship was built of Aleppo pine, Lebanon cedar, and oak. The *Santa Cruz* is built of teak, another fine wood. So it's no wonder that she's still in pretty fair shape.'

Upon Blake's orders, Captain Ike sailed the ship the half mile or so to the point where the little red flag bobbed up and down on the waves. There he dropped anchor.

Blake, Hal, Roger and Skink strapped on their aqualungs. It took longer than usual, for their fingers were crazy with excitement. Beneath them lay a ship loaded with treasure worth perhaps half a million. It was enough to make your fingers fumble over the buckles.

Captain Ike drew Blake to one side.

'Are you going to let Inkham go poking around that wreck?'

Blake was surprised. 'Why not?'

'I don't trust him.'

'I don't either. But I fail to see what harm he can do.'

'Don't you remember what he said — that if you found any treasure he would take it for himself?'

Blake laughed. 'Now be sensible, Captain. How can he make off with any treasure? You don't suppose he's going to swim away with it? And he has no ship. What can he do?'

'I don't know,' the captain admitted. 'But I bet he knows. He's a sly one. I don't trust him. He's threatened to grab the treasure and kill you. I don't think he was fooling. If I had my way I'd keep him locked up in the storeroom until we can put him on the plane.'

'I think his bark is worse than his bite,' Blake said. 'Don't worry, Captain. We'll keep an eye on him. We won't let him walk off with the *Santa Cruz*.' He grinned and hoped to see a responsive smile on the face of the kindly old sea dog. But Captain Ike only grunted and went off shaking his head.

The four explorers, with waterproof electric torches hooked to their belts, swam down, following the buoy line. At first they could see nothing below. Then the stumps of three masts appeared. They were bare, the rigging and sails having long since crumbled away. Then two strange towers could be seen and, last of all, the deck that connected them.

On their previous descent Blake and Hal had landed on the bottom beside the ship and had gone around it. Now Blake made straight for the deck and the others followed. In a moment they were standing on planks that had known no foot for three centuries.

The deck was covered with algae, sponges,

hydrozoa and gorgonias. Swarms of fish drifted about, for water creatures of all kinds seem to have a passionate affection for old wrecks. The bulwarks were high and a full three feet thick and pierced with holes for cannon. The cannon themselves lay on the deck, heavily upholstered in seaweed and coral.

Roger stooped to look into the mouth of one of the cannon but was drawn away by Hal who knew that cavities of this sort were exactly the kind of home preferred by the octopus.

What made Hal doubly suspicious of these cannon was the pile of stones and coral blocks in front of the mouth of each, almost closing the entrance. Those piles could not have fallen in place so neatly — they must have been put there by someone or something. And he knew that it was the habit of the octopus to retreat into a hole, then draw stones up to cover the opening, all except a gap just large enough for a tentacle to shoot out and seize its prey. Then the octopus would emerge to do battle, tumbling the stones out of its way.

With a quick movement of his hand Hal caught one of the pipefish that swam lazily around him. Holding one end of the long slender body, no bigger than a walking-stick, he dangled the other end in front of a cannon mouth. For a moment nothing happened. Then a tentacle shot out, seized the fish, and attempted to draw it into the hole. Hal hung on. After futile efforts to bring the fish to it, the octopus came to the fish, darting out of the cannon's mouth and flinging all eight tentacles around its victim. Hal thought it was time to let go and step out of the way.

He watched the octopus consume its prey, then slink back into the cannon and draw the stones up to the entrance.

Hal found himself alone. While he had been studying this little drama the others had moved aft. He wondered how he could spend time over a couple of animals when there was half a million dollars in treasure under his feet. Perhaps after all he was more a scientist than a treasure hunter.

He followed the others. They were approaching one of the two towers that loomed up at either end of the ship. These the old mariners called castles, and they did indeed look like castles. The one at the bow was three stories high with many windows and rich ornament. The stern castle was even finer, broader and taller, rising to four stories. The forward castle, probably used by the crew, was plain as compared with the magnificent stern castle where the officers and passengers had been quartered. On either side of the stern castle rose a splendid bronze lantern that any museum would value as a priceless relic.

The door from the stern castle to the gun deck had fallen away. The aqualungers entered the dark interior and turned on their torches. There was a scraping, swishing sound as dozens of small octopuses retreated into the corners, their angry eyes fixed on the intruders.

Staying close together to defend each other in case of attack, the men moved farther into the large room. In its centre stood a long heavy table securely bolted to the floor. The walls were lined with cabinets with leaded glass doors. With some effort Blake pulled open a door and would have gasped, if he could have done so without losing his mouthpiece, when he saw the silver and gold and ceramic platters, plates, goblets, cups, beakers, pitchers and bowls. Even if they found nothing else in the ship, here was fabulous treasure.

Dr Blake took out one of the platters and, having no polishing rag, rubbed it on the seat of his bathing trunks. Instead of the grey film that had covered the dish there now appeared a magnificent design of knights on horseback done in what appeared to be yellow gold, white gold and gunmetal.

Skink pressed forward and ran his fingers over the design. His hand was cramped in a curious way so that it looked like the claw of a bird. But he did not object when Dr Blake replaced the platter in the cabinet.

They climbed an ancient stairway, pausing now and then to give the frightened octopus population time to move out of the way.

Some of the beasts walked off daintily on the tips of their tentacles, while others shot away by jet propulsion.

The second floor and third floor seemed to be given over to individual cabins. The doors were closed and the explorers did not try to wrench them open but left them for later attention. They went on up to the fourth storey.

Here they came out into a single large room, magnificent in its proportions and decoration, and surrounded by small, cunningly designed windows clouded on the outside by marine growth. This may have been the captain's cabin or, when the governor was on board, was doubtless assigned to him.

Skink suddenly started back in terror. The others turned their torches in his direction and found him staring at — could they believe their eyes? — a man in full armour seated in a great chair.

He sat at ease and seemed to be very much alive although one could not see his face behind the visor of his helmet. He did not rise to greet his visitors but seemed to be studying them with sardonic humour. Perhaps he was enjoying their surprise at finding him there, a Spanish Rip Van Winkle three hundred years old, and to all appearances as healthy and happy as when he had last seen the light.

Skink, who was a bundle of superstitions, began to shake and had to sit down on a chest. The others tried to put on a bold front — but even they started back in fear when the old don began to smoke his pipe. What else but a pipe or cigar insidc that helmet could cause the fine column of black smoke that was coming out through the visor!

All that was needed now to terrify the beholders was for the man to move. And this he presently did.

A smile broke over the face of the helmet; one corner of the mouth went up in a one-sided grin. On up it went until the appearance was fantastic. Now it was like one prong of a moustache poking its way out of the helmet.

Hal started forward and brought his beam to

play full upon it. It was the tentacle of a small octopus that had made the helmet its home. Doubtless the black smoke had come from the same creature.

The tentacle waved about languidly as if it were the tip of a long moustache being stroked by invisible fingers. Then it slowly retreated again inside the helmet.

Hal's foot struck something on the floor. He turned his light downward and discovered two more mailed figures lying on the deck. One of them was cramped as if he had died in agony. Beside each figure lay a sword, its outline clearly defined through a film of slime.

Men did not ordinarily wear armour on shipboard – except in war, or an attack by pirates. Or to fight a duel. That seemed to be the explanation in this case.

But why did the man in the chair also wear armour? Perhaps he was to fight the winner. The ship had sunk just in time to save him the trouble.

Whatever the solution of this mystery, one thing was clear – these three magnificent suits of ancient armour would be prized possessions of the Metropolitan Museum. At least this was clear to three of the onlookers. The fourth, Skink, may have had other thoughts.

Finding that the three apparitions were neither living men nor ghosts, he crept forward to run his covetous fingers over the gold inlay in the steel helmet of one of the fallen warriors, over the neck guard and shoulder pauldrons, the handsomely engraved breastplate, the bulging elbow cops, the richly embossed gauntlets, the greaves that had encased the legs and the footgear made of narrow plates of flexible steel.

With the back of his sheath knife, Dr Blake pried open one of the several chests in the room. It was full of marble and porcelain statuettes of great beauty. Another contained two gold peacocks set with jewels. Another contained nothing at all but a little deposit on the bottom, all that remained of fine fabrics, perhaps tapestries, perhaps clothing.

At one side of the room was a large bed and at its foot was — of all things! — a silver bathtub.

Dr Blake was startled to see an almost naked man stretched out in it — until he noticed who the man was. It was mischievous Roger who hopped up, laughing so hard that he almost lost his air.

What a task it must have been to fill this tub! It was quite innocent of plumbing and must have been supplied by means of buckets laboriously carried up three flights of stairs. Then the ship had found a way to avoid all this bother. She had only to sink to the bottom of the sea and the silver bathtub would remain full for ever without trouble to anyone.

Blake led the way down to the gun deck and found a companionway descending into the hold. Here there were larger members of the octopus family, but so long as they were not trapped or cornered or otherwise annoyed they did nothing but glare at the intruders. There were also many big fish that had come in through broken hatches.

The hold was filled with household goods and treasures of fine design and workmanship. Some were Philippine in origin, some Chinese, some Indian, but most of them had evidently been imported from Spain to outfit the governor's mansion in Manila and had, after his resignation, been on their way back to Spain when the ship sank. There were bronze and stone lanterns, crystal chandeliers, marble statues, large garden vases, a bronze sundial, more efficient timepieces in the form of highly ornamented clocks and bulky watches of the old style with hour hand only, their dials being covered with enamel. There were chests and chests and chests of miscellaneous articles, swords, rings, buckles, chains, necklaces, unset jewels, slugs of pieces of eight, golden doubloons, gold and silver bullion and coins.

At one place the hull beneath their feet was wrenched open and the sand of the sea bottom pushed through. That told the story of the sinking of the *Santa Cruz*. Unwieldy because of her heavy castles, she had been twisted by the storm until the strakes of her hull parted and let in the sea. A chest had broken and tumbled its load of gold pieces into the hole. Skink moved to pick up some of it but Blake gestured that it should be left as it was.

Skink was puffing so hard with excitement that he exhausted his main air supply and had to press the lever that gave him his last five-minute reserve. Blake, realizing that all the tanks must be nearly empty, signalled the ascent. The four masked figures soared up through an open hatch, rose to the broken tops of the masts, hung there for a few minutes to adjust to the change in pressure, then continued to the surface and climbed on board the *Lively Lady*.

Roger was bursting with a long-suppressed question.

'Why didn't we find any men on that ship, except those three?'

'We didn't find any men,' Blake answered.

'But the three in the cabin . . .'

'Those were just empty suits of armour.'

'But there must have been bodies inside, or at least skeletons.'

'When we open those suits I don't expect to find a human fragment as big as the knuckle of your little finger. The flesh was probably eaten within a few hours by fish and starfish and crustaceans. And within a few weeks the worms and bacteria did away with the bones. Metal and stone and some kinds of wood will last, but not bone.'

It seemed a grim thought to Roger that men, who think themselves so wonderful, should vanish so quickly while metal, stone and wood last for centuries.

'We're not so very important, are we?' he said, a little mournfully.

Blake laughed. 'Are you just finding that out? Now, let's get to work. We're not going to take a thing from that wreck until we have photographed it from stem to stern, inside and outside. Then we'll begin removing the cargo.'

'Do we have to get a permit from the Trusteeship?' Hal inquired.

'That's all arranged. There won't be any government tax on the property so long as it goes to the museum. And that's where it's all going to go.'

Blake heard a snort behind him and turned to see Skink. Skink at once wiped the smirk off his face and said nothing.

'We want pictures of everything,' Blake went on, 'just as it is – the armed men, the chests, the cargo – in black and white, colour, and movies.'

'How about a few paintings?' Skink suggested.

'That would be very interesting. Why don't you try it?'

The aqualungs were refilled, the cameras, flash equipment and painting material assembled, and Blake, Hal and Skink descended to the wreck.

Blake went inside and proceeded to photograph, with the help of flash, the cargo and the dramatic scene in the upper cabin. He also made notes on what he saw – and just as he had been astonished at the sight of the man in the big chair, the man in the chair should have been surprised to see this curious creature with a mask on his face and a tank on his back, calmly seated on a chest and writing on a slate with a slate pencil.

The water dimmed the effect, but when the slate dried off the writing would stand out clear and white. Blake had learned this method, as well as the trick of writing on a zinc pad with a lead pencil, from William Beebe. A third method was to write with a graphite stick on sheets of sandpapered xylonite, a waterproof substance rather like celluloid.

Such notes made on the spot were necessary to an accurate and scientific record, because after rising to the surface it was so easy to forget exact details.

Hal, outside the ship, was taking general views of the wreck as it lay in the sand, the gun deck and bulwarks, and the two picturesque castles. He was especially interested in the long beakhead of the ship, heavily carved with animals, monograms, crowns, serpents and floral ornaments and terminating in a splendid bronze figurehead of Neptune rising from the sea. He could already see that striking work of art in an alcove of its own in the Metropolitan Museum. Perhaps below it there would be a caption referring to the expedition of the *Lively Lady* and naming the scientists who had discovered the *Santa Cruz*.

He saw that someone else was much interested in the figurehead. Skink was painting a picture of it. Seated on a coral block, with a board-backed canvas on his knees, he was having unexpected troubles. The board wanted constantly to jump away from him and soar to the surface. In trying to keep it down he lost hold of his brush and it immediately 'fell up' out of sight. Considerably annoyed, he pulled another brush from his belt. He squeezed colour from tubes on his palette and was astonished when a tube labelled red gave out green paint and one marked yellow emitted grey. He knew from experience that red blood looks green at a depth of sixty feet, but had not realized that his paint would be affected in the same way.

Small fish swarmed between him and his canvas so that he could hardly see what he was doing. They were very curious to know what was going on

and while some of them nosed their way over the picture, smearing the lines, others pressed their noses against the glass of his mask.

He was disconcerted to find that as fast as he put colours on his palette, they disappeared, and he had to squeeze out more. Then he observed that the fish were eating his colours. They evidently had a good appetite for oil paints.

Nevertheless he kept at it and made a picture. It took many colours to paint the rainbow tints of the corals and seaweeds and sponges and gorgeous tropical fish that surrounded the old figurehead which was itself covered with marine growth of every tint and shade.

At last he put his head on one side, inspected the finished work, and told himself that he had created a masterpiece.

Blake appeared on the gun deck and signalled the others to come up. They did so and he led them into the stern castle. There a surprise awaited them. The table had been set for lunch.

Blake had brought down with him a tin lunch box in which he had placed three small cans of sausages and three bottles of Coca Cola. These he had now placed on the table and gestured his companions to sit down on the long bench. This they did, but waited in some bewilderment to see how Blake proposed to eat and drink under water.

As a matter of fact, Blake himself had never tried it before, but he had seen the divers at Weekiwachee Springs, Florida, calmly chew celery and drink pop at a depth of thirty feet. He had no celery, but sausages ought to serve instead.

With the point of his knife he cut open his can and extracted a sausage. He removed the aqualung mouthpiece from his mouth. For as long as it would take to eat this sausage, he would have to do without air.

He puckered his lips, pressed the end of the sausage against them and forced it slowly in, taking care not to leave any opening on either side of the sausage where water could enter. He pushed the sausage all the way in and the lips closed over it. Then he munched contentedly while the smile of a

Cheshire cat came over his face. He replaced his mouthpiece and breathed.

Hal and Skink followed his example, and the process was repeated until all the sausages were gone. But there still remained the puzzling problem of how to drink a bottle of Coca Cola ten fathoms beneath the sea.

When Dr Blake pried off the cap of his bottle a strange thing happened. Since the pressure outside was so much greater than that inside the bottle, sea water immediately entered and compressed the contents. But a little sea water did no harm, and Dr Blake pressed the mouth of the bottle to his lips.

By breathing out into the bottle he displaced the contents which thereupon flowed into his mouth. He drained the bottle. When he took it from his lips the sea water filled it with a sudden thud. Hal and Skink faithfully followed the same procedure.

After a little more work they surfaced and boarded the *Lively Lady*.

'You're just in time for lunch,' cried Roger.

'Thanks,' Blake answered. 'We've had lunch!'

But it took very little pressure to persuade them to sit down to some of Omo's good cooking. However, before they could eat, Skink must show them his masterpiece.

With a flourish, he uncovered his canvas.

Everyone tried to be polite, but it was very difficult not to laugh. Roger turned red and nearly choked. The captain suddenly remembered something he had to do on deck.

It was really a terrible mess. Every colour swore at all the rest. Not one of the colours was what it had seemed to be at the bottom of the sea. Because of the peculiar way in which water absorbs light, nothing was now what it had been under the filter of ten fathoms of blue water.

All Skink could do was to say dolefully 'Well, if you'll just come down to the bottom and look at it, you'll see it's pretty good.'

But since no one was particularly interested in a picture that had to be viewed at the bottom of the sea, his offer was not accepted.

After the second lunch everyone took a siesta — everyone but Skink. He excused himself with 'I want to go down and make another try at that picture.'

When he came back an hour later with a blank canvas, Hal asked him what had gone wrong.

'Oh, I had bad luck,' Skink said. 'I had the picture almost finished when a school of about a hundred parrotfish came along and ate every bit of paint off the canvas.'

Hal studied Skink's sly face. It just might be true, but it was a pretty big story.

Could it be that Skink had not been painting at all? But what else could he have been doing? He surely could not steal anything from the wreck. He had no place to hide it. He wore nothing but brief bathing trunks that certainly afforded no hiding-place for a suit of armour or a chest of bullion.

Hal dismissed his suspicions and turned to his laboratory work. But he was still uneasy and finally decided to go down and take another look at the wreck.

As he sank slowly towards the bottom he thought

he saw at a considerable distance a roundish black object hovering in the blue. It looked like a small submarine, but of course it couldn't be that. It must be a large fish, perhaps a black manta.

He thought no more of it and landed on the deck of the *Santa Cruz*. Entering the stern castle, he was surprised to see the doors of several of the wall cabinets hanging open. Inside, there was nothing. The fine platters, plates, goblets and all had disappeared.

His heart pounding with excitement, he half-walked, half-swam up the stairs to the upper cabin. The armoured man had risen from the chair and departed. The two men who had been lying on the floor had vanished.

He went down into the hold. Here everything seemed to be as it had been. The thief, or thieves, had not had time to make away with all of the cargo. But they had certainly made a good beginning.

Had they taken the splendid bronze figurehead? He rose through the hatch and swam to the bow. The figurehead was gone.

Some instinct told him that the stolen treasures must be near by. He descended to the sand and circled the ship. There were large trees of elkhorn coral and much small growth, but no hiding-place for loot.

He systematically made another circle, twenty feet farther out. Then another. And another.

At last, about three hundred feet to port of the vessel, he came upon bottom of a different character. Here there were huge igneous rocks left by some ancient volcanic upheaval. Between the great boulders were cracks and caves and these he explored carefully, watching out for the moray eel and the octopus which love such retreats.

In the heart of the labyrinth he came upon a grotto penetrating so far back under the rocks that he had to use his torch. He had a sudden shock when the beam of light picked out a man standing quietly near the back wall of the cave. Then he saw that it was the life-size Neptune, the figurehead of the *Santa Cruz*.

Around it were all the other stolen articles, including the silver and gold tableware and the three suits of armour.

There was only one person who could have done this — Skink. Hal felt the blood boiling in his veins. He would go up and have it out with Skink. He would show him up for the scoundrel he really was.

First he would take these things back to the wreck. On second thoughts — no — he would leave them right here. He would bring Skink down into this cave, face to face with the stolen goods. Then the fellow could not deny his guilt. He would stand convicted as a thief and they would deal with him as a thief.

Full of grim resolve, Hal carefully noted the position of the cave and then returned to the *Lively Lady*. As he climbed on board Blake said,

'How's everything down below?'

'The ship is still there,' Hal muttered.

'Good,' laughed Blake. 'That's one comfort. Nobody is likely to carry off the *Santa Cruz*.'

'Not all in one piece,' Hal said.

Blake looked puzzled. 'Now what do you mean by that?'

'Just that we have a crook on board and he's been stealing stuff from the wreck.'

Skink, lying on the deck writing his notes, looked up questioningly.

'That's a serious charge,' Blake said. 'What is missing from the wreck?'

'Gold and silver plate, the three suits of armour, and the figurehead.'

Blake studied Hal. 'You must have made some mistake. Are you sure you feel all right? Rapture of the depths sometimes does funny things.'

'I'm not depth-drunk,' Hal insisted. 'The things are gone from the ship. And I know where they are.'

Skink looked up again, his jaw dropping open.

'I found them in a cave where Skink put them.'

Skink leaped to his feet and advanced upon Hal. 'Do I understand that you're accusing me?'

'I hope you understand it,' Hal said, 'because that's just what I'm doing.'

Skink prepared to swing on his enemy, but Blake pushed him aside. 'There's an easy way to check on all this,' he said. 'We'll go down and take a look in that cave.'

'Good idea!' growled Skink. 'Nothing would suit me better. Let's go – as soon as I fill my tank.'

It was true that all the tanks needed to be filled. Hal chafed at the delay – yet he supposed it didn't really matter whether they went at once or a bit later.

The compressor was started, and the work of

recharging the cylinders began. Skink succeeded in giving the impression that he could not wait to get below the surface and disprove the accusation made by Hal against him. He was impatient with the machine.

'I think it has some worn bearings and piston slap,' he said. 'Let me at it – I think I can speed it up a bit.'

Hal did not trust his sincerity and was not surprised when instead of speeding up the compressor Skink took the thing apart and kept it dismantled for more than half an hour. When the machine was reassembled it worked no faster than before. The better part of another hour passed before the tanks were filled.

In the meantime Skink's eyes kept searching the sea. Hal, suspicious, followed his gaze, but there was nothing to look at — the sea for miles around was a perfect blank. He did finally detect a black object projecting above the surface and moving in the direction of the island, but it appeared to be only the fin of a large fish. It rounded a bend of the island and disappeared behind the coconut palms.

'All right, let's get along,' cried Skink. 'I can't wait to show up this four-flusher.'

Aqualunged and masked, Blake, Hal, Skink and Roger descended to the wreck and then, led by Hal, swam out three hundred feet to the labyrinth of rocks. Hal conducted them unerringly through the twisting passages to the mouth of the cave.

The interior was completely dark. Blake would have turned on his torch but Hal restrained him.

He took Blake and Skink in to a point where, when the light went on, they would be face to face with the stolen treasures. He wanted to see just how both of them would act when the evidence of Skink's guilt loomed up before them.

Like a stage manager striving for dramatic effects, Hal made them wait a few moments in the dark so that the scene would be all the more startling when the lights went on.

After an impressive pause, he signalled them to use their torches by clicking on his own.

Everything was suddenly bathed in blinding light. Every detail of the rocky wall, ceiling, floor, every crack and corner, was clearly revealed.

Hal could hardly believe his eyes, and yet there was not the least doubt about it. . . .

The cave was empty.

11
Night Dive

Blake and Skink turned to fix questioning eyes on Hal. He pressed through between them and went to the end of the cave.

He ran his fingers over the wall as if to make sure that it was solid – not just a screen to hide the treasure. He felt into every crack and hollow. He explored the bottom on the chance that there might be a hole through which the treasure had dropped.

He knew that he was making himself quite ridiculous. Blake would be more sure than ever that he was affected with nitrogen fever, drunkness of the depths, that caused divers to see strange visions, things that just weren't there.

Had he really seen the treasure in this cave? His mind was confused and bewildered. Perhaps he had spent too much time at this depth and the pressure was upsetting him.

Or perhaps this was not the right cave! It would be easy to make a mistake among the many caverns of this wilderness of rocks. That was it – he was just in the wrong cave.

He went outside and looked at the entrance again, and the surrounding boulders. He saw all the landmarks that he had committed to memory – that huge brain coral, the elkhorn tree shaped like a cross, the tall rock leaning forward that he had thought looked like a giant old woman. He was sure this was the place.

The others were rising to the surface. He went back into the grotto. He had a dim hope that by some magic the treasure would reappear. It did not.

He joined the others on the deck of the *Lively Lady* and found himself apologizing to Blake. 'Sorry I took you on a wild-goose chase. But I could have sworn . . .'

Blake was very patient.

'I know. It's a strange world down there and if a fellow stays too long he gets some of the strangeness into his head. What you need is to rest on deck. No more diving for you today.'

'Perhaps you're right,' Hal said wearily and stretched out on the deck.

'Before you get too comfortable,' Skink said acidly, 'you can apologize to me. You said something about my stealing that stuff – remember?'

'I'll apologize fast enough if I'm wrong,' Hal

answered, 'but I'm still not convinced. There's something queer about all this.'

'I'll say there's something queer,' Skink said contemptuously, '— your head.'

Hal did not answer.

Roger, with the kink that thinking always put in his forehead, sat studying his brother. It wasn't like Hal to go nuts. His head was like a good watch — you could always depend upon it. If he said he saw the treasure in that cave, he saw it. Suddenly the kink disappeared from Roger's forehead and he called to Dr Blake:

'There's one thing we forgot.'

'What's that?'

'To look in the wreck and see if those things are really gone.'

'Of course they're not gone!' Dr Blake was having difficulty in keeping his temper. His patience was sorely tried. 'Now, look at it sensibly. How could anyone take anything from that wreck and make away with it under water? He could put it in a cave, but what good would that do him? He couldn't take it away without a ship. And if it stayed in the cave very long we would discover it. Anyhow, we found it wasn't in the cave.'

'Perhaps Skink had already removed it.'

'How could he? You forgot that he was on deck every minute from the time Hal reported the treasure was in the cave till the time we went down and found it wasn't there. How could he have removed it — by magic?'

Roger shook his head. It was too deep for him.

'Another thing —' Skink put in, 'your brother

claims that the figurehead was in the cave. That thing was life size and solid bronze. It must have weighed a quarter of a ton. You flatter me if you think I'm strong enough to pull that off the ship and carry it three hundred feet.'

A broad smile of self-satisfaction spread over Skink's face. Let the runt answer that one.

Roger did. 'I could have carried it off myself. I took a look at that figurehead. It wasn't fastened to the bow — it had broken off, and was just held in position by the rocks. And as for the weight — I don't think that thing would weigh more than three hundred pounds above water and at a depth of ten fathoms its weight would be cut to a hundred pounds — just a good heavy load for one man.' He turned to Blake. 'Isn't that right?'

Blake nodded. 'But you still haven't explained how the stuff disappeared from the cave, if it was ever there.'

Hal resented this remark. 'It was there,' he insisted. His mind was clearing now. He was sure he had not dreamed all this. 'And if you'll come down with me again you'll find it isn't in the wreck.'

Skink hastily objected. 'You won't drag us down there on any more fool's errands.'

Hal ignored Skink and addressed himself to Dr Blake. 'What can we lose by going down? Even if there's only the faintest chance that a thief is working on the cargo, isn't it important for us to find out about it?'

Blake sighed deeply. 'You win! Just to satisfy you, we'll go down.'

'How about making it first thing tomorrow

morning?' Skink suggested. 'It's getting dark and I think Omo has supper ready.'

Blake wavered. The good smell of hot food came up from the cabin. Then he saw something in Skink's face that made him say, 'No, we'll go now.'

They sank through the darkening sea. The torches were clicked on even before they reached the deck. They entered the stern castle.

The doors of the cabinets hung open and the cabinets were empty. It was Dr Blake's turn to show excitement. He went to the cabinets, explored them, searched the room, even peered under the table, then stood staring at Skink with a look that made that gentleman squirm.

Turning, Blake led the way up the stairs to the master cabin. Their torches lit up the great room and several fish and an octopus swam out of reach. The beam of Dr Blake's torch picked out the great chair. The figure that had made itself at ease there for three hundred years was gone. The fallen warriors had also disappeared.

They went down and out and around the forward castle to the prow. There was no bronze Neptune.

Back on the *Lively Lady* Blake tore out his mouthpiece and gave vent to the anger and dismay that he would have liked to express at the bottom of the sea. After he had blown off he turned to Skink.

'Inkham, what do you know about all this?'

'Nothing at all,' said Skink blandly. 'Apparently Hunt is the only one who has any information on the subject. It seems that he went down, and the stuff disappeared. He says I took it. Isn't it more likely that he did?'

'Don't be ridiculous!' exploded Blake. But he was deeply puzzled. He knew that Hal Hunt was not capable of such an act, and that Inkham was. But Inkham seemed to be in the clear. Then was someone else the thief? The captain? Omo? Impossible. How could any outside person have done the job? They were a hundred and fifty miles from anywhere.

'Captain, have you sighted any ship today?'

'Nary a one — and don't expect to. We're away outside the lanes.'

Blake cudgelled his brain. 'The invisible man — who could he be? And where could he take the stuff? The nearest place is this island. We'll investigate that tomorrow morning. In the meantime, we'll take no more chances. We'll keep a guard on that wreck day and night. The watches will be one hour long. I'll go first, then Hal, then Roger, then Omo. Then we'll repeat.'

'How about me?' inquired Skink.

'We'll let you get a good night's rest.'

Skink glowered fiercely but said nothing. He went down to supper and the others followed. Blake ate lightly, for it was not well to dive after a heavy meal. Then he sank into the black sea, trailing light from his torch like a departing comet.

After an hour he was back, reporting all quiet, except that hundreds of strange fish never seen by day had come in from deep water and were milling around the wreck.

'You'll have plenty of company,' he told Hal.

'Every octopus in the ship has come out of its hole and is spending the evening on deck.'

While Hal was below Roger tried to catch a nap, but the unhappy anticipation of having to spend an hour at the bottom of the sea at night kept him fully awake. When it came to his turn, he would have swapped it for a brass farthing. He made sure that his knife was sharp and took along a shark billy as well.

He climbed down the ladder to the lowest rung and stood there for a full minute, mustering up courage to go in. There was no moon but the sky was full of stars. The cold night breeze made him shiver and think of his bunk. The ship's smells of sails and planks and engine oil seemed mighty good. Was it really necessary to keep guard on the wreck?

Hal leaned over the rail. 'If you don't fancy going down,' he said, 'I can take another turn.'

Roger didn't know whether to be grateful or offended. He released his hold on the ladder and swam down.

If the sky had been full of stars, so was the sea. He could imagine himself swimming through the Milky Way. Millions of phosphorescent lights flashed on and off. The lights were sometimes single, sometimes in rows, sometimes in star shapes, sometimes in circles. They were red, yellow, green, blue, lavender.

He tried to imagine what strange shapes, what fish, serpents, monsters, lay behind them. He turned on his flashlight.

It made a conical beam of light in the water, but

outside of the beam everything was darker and more mysterious than before. He felt as if jaws were about to close on him from behind. He swung around, playing the beam in every direction. This only blinded him the more.

It took nerve to turn off the beam. At first he could not see a thing, not even the phosphorescence. His eyes gradually became adjusted to the millions of underwater traffic lights, and he could even make out the shapes behind them.

Some of the illuminated creatures lit up others. A school of glowing jellyfish cast a ghostly light over a big grouper that seemed to be studying Roger with interest while opening and closing its mouth as if it were saying, 'Oh, brother!' A scurry of shrimp lighted the stump of a mast of the *Santa Cruz*.

He followed the mast down to the deck.

A large fish swam by leaving a trail of phosphorescence behind it. It illuminated hundreds of small octopuses waltzing over the deck on the tips of their tentacles. Roger thought it just as well not to settle on the deck but to float ten or twenty feet above it.

Even here the eight-legged merrymakers did not leave him alone but occasionally shot by him like comets, all tentacles laid close and straight to make the body perfectly streamlined.

The rock and coral formations of the undersea landscape were picked out in lights and the big fifteen-foot elkhorns were as fantastic as Joshua trees. Everything was on the move. The sea worms, sea urchins and starfish trotted about with an alacrity that one would never suspect, having seen them only by day. The conger eels and morays that play a waiting game all day long had come out of their holes and were actively searching for food. They plunged at any small flashing thing. Roger was glad that he had rubber fins on his feet, and he kept his hands close against his body. A flick of his fins now and then was all that was necessary to keep him suspended.

In the darkness the sounds of the underworld seemed louder than they had ever been by day. Did he hear someone climbing up on to the deck? No, it was just a large fish scratching itself against the wreck to scrape off some parasites. There were grinding sounds that perhaps came from parrotfish crunching coral with their horny beaks.

He knew that many fish had been named because of the sounds they make, and probably the grunts he heard came from the grunts and the croaks from the croakers and the squealing sounds from the pig fishes.

As a matter of fact, the supposedly silent sea is full of sounds and Roger heard only a few of them that night. The schoolmaster rumbles as if delivering a lecture. The oldwife chirps and chatters. The drumfish drums, the porpoise snorts and the singing fish produces musical notes much in the manner of a cricket or grasshopper.

But to Roger every sound he heard was made by the invisible man coming on board to steal more of the treasures of the *Santa Cruz*. A dozen times he heard him creeping stealthily over the bulwarks – but each time he proved to be a fish or octopus.

He did not light his torch again since that would

instantly reveal his position and he might be attacked from behind. He kept close to the mast so that its shadow would blend with his own. Even so, he did not feel at all comfortable. It was enough to be startled every moment or so by the whizz of a passing octopus or the curiosity of a fish poking his flesh with its nose, without the apprehension that the invisible man, or men, might gang up on him. With so rich a store almost within their grasp, they would not stop at murder.

He suspected Skink, but Skink could not have engineered the thing alone. He could have taken the treasure to the cave, but who took it from the cave? Not Skink — he was on deck. It could hardly be the work of an octopus or any other marine creature. It must be the work of humans — if people who lived under the sea could be called humans. Could they possibly be some strange breed of people that breathed by means of gills rather than lungs? His fancy played with this idea.

He was jolted out of his reverie by a sudden burst of light coming from the open hatch leading into the hold. There was someone in the hold of that wreck! With the help of his flashlight he must be exploring the treasure, trying to decide what to steal next.

Roger must go to get help. But it would take many minutes for the others to get on their bathing trunks, masks, aqualungs, belts and fins and come down. By that time the thief might have departed with a load of loot.

Perhaps the intruder could be frightened away. Roger took off his weighted belt, and struck the lead discs, each weighing a pound, repeatedly against the mast. The dull thudding sound was carried down into the ship and echoed through the hollow spaces of the wreck. If anyone was down there he must be frightened by this booming sound that could not possibly come from any fish.

But the light in the hold was not turned off and no one rose through the hatch. The fellow must be deaf! Well, perhaps a beam from Roger's own torch would startle him.

Roger shot a ray of light into the hold. He moved it here and there. He turned it on and off, on and off. It was a strong ray, strong enough to cut through the glow between decks. It had no effect.

He noticed a peculiar thing about the light in the hold. It appeared to be constantly changing, throbbing, brightening, dimming, and brightening again. It was hard to imagine the light from a torch behaving in this manner.

He flicked his fins upwards and sank close to the open hatch. After giving the octopuses time to scatter, he lowered himself to the deck, gripped the coaming of the hatch, and put his head in far enough so that he could see all of the interior of the hold.

A terrifying sight met his eyes. Roger had never believed in sea serpents. Yet what could this be if not a sea serpent, this snake-like monster tearing about madly from one end of the hold to the other, thrashing the thousands of small fish and other organisms so that their light cells were agitated to their greatest activity? It was not round like a snake but flattened, almost like an immense ribbon, silvery-sided, with a small mouth and deep-set terrible eyes. But its most amazing feature was a flaming red mane like the mane of a horse that stood straight up from its head and neck. This waving, billowing mane did a sort of dance of fire in the unearthly glow cast upon it from all sides as the creature charged furiously back and forth. Two long spikes that looked as sharp as daggers projected from the back of the head.

It must have come from great depths to spend the night in the upper waters as many deep-sea fish do. It had blundered into the hold and was now frantically seeking a way of escape.

What a prize for a zoo! Roger had never seen anything like it in any aquarium. But how could he take it? Even if he had a rope, and he had not, he dared not venture into the hold. The creature's mouth was small, but the teeth looked very efficient and the two daggers were not to be forgotten. A lash from that tail would knock a man senseless.

He struggled with the ancient iron hatch cover. One edge of it was glued to the deck with barnacles. He systematically prised these loose with his knife. The hatch cover was very heavy, though only a fraction of what it would have weighed in the air. He finally inched it up over the hatch.

Then he returned to the *Lively Lady*, shook Hal and Dr Blake out of their sleep and told them what he had found. Without waiting for them to accompany him, he seized a net and returned to the wreck.

He dragged off the hatch cover and spread the net to cover the opening. He drew it down snugly

over the edge of the coaming all around and tied it in place.

Sooner or later the sea serpent, or whatever it was, would discover this exit and dash out, only to be caught in the net. But then what? All alone, he could never wangle it up to the ship. He wished those fellows would hurry.

They came at last and Blake seemed to approve of Roger's plans. He gazed with amazement at the raging, red-maned serpent shooting like a jagged bolt of lightning from end to end of the hold. Several times the red thunderbolt barely missed the net. Blake untied it so that it would come free when necessary and he with Hal and Roger held the edges.

The serpent suddenly exploded through the hatch, lifting the net and the three men hanging on to it many feet above the deck. They hastily closed in and the raging monster was trapped.

Still it squirmed, struck, flailed, bulging the net into all sorts of fantastic shapes. It plunged out at Roger and its sharp teeth grazed his arm. The flashing daggers and the whipping tail had to be watched with care.

Reaching the side of the *Lively Lady*, Blake called to the captain to lower a line. The line was made fast to the net and the violent passenger was hoisted aboard and lowered into a tank. There it was released from the net and proceeded to turn the water into foam.

'An oarfish!' Blake cried. 'Twenty feet long if it's an inch! It's a young one. If it lives, and that's a question, it might grow to forty feet.'

'It looks like a sea serpent,' Roger said. To him the name oarfish seemed pretty dull.

'It *is* a sea serpent. Or at least what is known among sailors as the sea serpent. It lives far down, but it sometimes comes to the surface, and who can blame anyone who sees that fiery head and twenty to forty feet of twisting body for calling it a sea serpent?'

'So it's really not a serpent?' Roger mourned.

'No, nor a snake or eel. It's a fish – called oarfish because it is flat like an oar. But you don't need to be disappointed. It's a wonderful catch, the best we've made yet. And I think you deserve a special favour. No more watches for you tonight.'

Roger did not refuse the favour. Gratefully he stripped off his diving gear, got into pyjamas, and slunk into his warm bunk.

12
The Man-eating Clam

Over a smooth sea sparkling under the low light of the early sun, the dinghy ploughed towards the island. On board were Blake, Hal, Roger and Skink.

Blake had considered leaving Inkham on the ship, but had decided that it was just as well to bring him along so that he could be kept under constant observation. He suspected the fellow, even though there was actually more evidence against Hal than against Inkham.

Perhaps neither one was guilty and it was all the work of some 'invisible man'. It was very perplexing.

Blake shut off the motor but did not run the boat up on to the beach. He stepped out into the shallows and said, 'Inkham and I will do this side of the island; you two explore the other side. Take the boat. When we get done here we'll walk across and join you. If you find the loot, whistle.'

The plan did not suit Hal. He hated to leave Blake alone with his worst enemy. Skink's threat that Blake would have a very bad accident, that he, Skink, would become boss of the expedition, and that if any treasure was found he would make it his own, stuck in Hal's mind. Perhaps the fellow was just bluffing — but perhaps he wasn't.

'Wouldn't it be better for us to stick together?' Hal suggested.

Blake was already wading to the beach. He turned and asked, 'Why?'

Hal mumbled, 'Just an idea.' He could hardly say that he was afraid the director was incapable of taking care of himself.

'We can do the job twice as fast if we split up,' Blake said. 'Come along, Inkham.'

The boat chugged away, its noise making a dwindling path through the morning stillness while Blake and Skink walked along the beach close to the trees and watched for the mark of a keel, the print of a foot, the ashes of a fire, an empty food can, a path through the underbrush, or any other sign of a recent landing on the island.

The palms and pandanus cast long-fingered shadows on the beach. Now and then a coconut fell with a thump. The trade wind was as refreshing as a cool drink and the sky was that deep solid blue found only in the South Seas and the desert. It was a glorious morning — the sort of morning when nothing unpleasant could possibly happen.

But Skink was thinking unpleasant thoughts. Now was his chance. Half a million dollars were at stake. A fellow would do a lot for half a million dollars.

But how would he go about it? It would be easy to slide up behind Blake and sink a knife between his shoulders. But that would only make new problems. If Blake disappeared, he, Skink, would be blamed. If the body were found, the knife wound would betray the killer.

Perhaps he could make away with Hal and Roger as well, then they could tell no tales. But that was too big a job. He knew from experience that Hal was not an easy victim, and Roger was almost as strong.

They came out on the shore of a small bay. At the back of the bay the land rose straight from the water in a sheer cliff several hundred feet high.

'They certainly couldn't have landed there,'

Blake said. 'No use of climbing over that thing to get around the bay. Suppose we swim across.'

'Why not?'

Their singlets, dungarees and canvas sneakers would get wet, but would soon dry again.

Blake walked to the edge of the water. 'Turn of the tide,' he said. 'Running in fast. But it's still low enough so that I believe we could almost walk across. Let's go.'

They waded in, up to their waists, up to their chests. The bottom was smooth, hard sand. The incoming tide pressed against their bodies and they had to lean seaward to resist it.

Skink collided with something big and solid. He thought it was a rock. Looking down, he saw that it was a giant clam. Its great shell had snapped shut, barely missing his fingers.

He was about to announce his discovery, then thought it better to say nothing. A new hope began to glimmer in his mind. When the *Lively Lady* had first neared the island they had noticed that the shallows abounded in giant clams. These monster mussels measured six feet across and weighed up to eight hundred pounds. They lay anchored to the bottom with their great jaws open and closed them promptly upon anything that came between them. The creature had no preference for human flesh but, since so many bathers had been caught in that terrible trap, it well deserved to be called the man-eating clam.

The water was getting a bit deeper now and it was necessary to swim. Skink pushed ahead of Blake and watched the bottom as he swam. He began to despair of seeing another man-eater. Then a huge one appeared directly ahead.

Skink swam cautiously over it, then stopped, barring Blake's path. 'Just want to rest a minute,' he remarked. Blake paused, allowing his feet to sink in seach of the bottom.

A contortion of pain came over his face and he cried, 'A shark! It's got my foot!' He drew his knife and put his head under water. Immediately he broke the surface again. 'It's not a shark. It's one of those devilish clams.'

He was chin deep in water. The tide was coming in. Within a few minutes, ten or fifteen at the most, mouth and nose would be submerged. In the meantime he was in great pain, but his voice was calm.

'Now, Inkham, I'll tell you just what to do. There's no use trying to cut it loose from the bottom. It's too heavy to lift. Your only chance is to get inside and cut the hinge.'

'That's a pretty large order, isn't it?'

There was something in Skink's voice that Blake didn't like. 'Yes, I suppose so, but it's the only way. Chip away the edge of the shell – make a hole big enough to get your hand in. Then reach all the way to the bottom and cut the adductor muscle.'

Skink wondered. Would he let the fool know now that his number was up, or would he keep him guessing a while? He would keep him guessing. Revenge was sweet – he would enjoy it as long as possible.

He drew his knife, dived, and pretended to chip at the edge of the shell. After nearly two minutes, he came up. He thought the water had risen a fraction of an inch higher on Blake's chin.

'Had to come up to breathe,' he explained.

'Of course.'

Blake waited patiently, his twisted face the only sign of his agony. Skink would have liked to hear him curse, rage, weep, go crazy with fear and pain. The scientist's composure was disappointing.

'Well,' Blake said, 'aren't you ready to go down again?'

'Sure.'

Skink went down, tinkered for as long as he could hold his breath, a good three minutes, and reappeared.

He breathed and blew for a few moments, then said, 'Sorry, I can't seem to make a dent in it.'

Blake's mouth was almost under water. 'That's all right,' he managed to say. 'You tried. There's one other way. Hack off my foot.'

Even the villainous Skink recoiled at this suggestion. 'I couldn't do that,' he said, and he meant it.

Poor fellow, thought Blake, it's not his fault if he's a coward. Aloud he said, 'Then I'll do it myself.' He drew his knife and submerged.

Skink was seized with a violent fit of trembling. He thought he was going to have a cramp and drown along with his enemy. He swam ashore and stood shaking on the beach. He dared not look back.

When he did, he saw nothing. He strained his eyes. For five minutes he looked, but still saw nothing. The tide was creeping over his feet.

He turned and walked off in a daze down the beach, the way they had come.

He had not meant to do that – or had he? Do what? After all, he had done nothing. If the fool had to blunder into an underwater bear trap, who was to blame but himself?

And what a fool! Up to the very end he insisted upon believing that Skink would help him. He had too much faith in human nature, that Blake. Skink tried to laugh, but couldn't. Somehow he felt awfully cheap, an odd sensation for him. He ought to feel on top of the world. His enemy was out of the way, and half a million dollars were as good as his. Why did his mouth feel dry and stale, as if he had smoked too many cigarettes?

Skink circled the island until he came upon Hal and Roger. He dropped wearily to the ground. His head ached and his nerves were jumping like a school of minnows.

'Where's Blake?' Hal asked.

'He went round the other way. I thought he'd be here by this time.'

Hal studied him. 'You look pretty well beat up. What happened?'

'Nothing. I just got a bit too much sun.'

'Well, there's some good shade under that bread-fruit. We've done this side, but we thought we'd go in and take a look at the lagoon shore. If Blake comes, call.'

Hal and Roger ploughed through underbrush, berry bushes, lantana, and a criss-cross of sago palms and pandanus towards the lagoon. They kept their eyes open, but it was not possible to see more than a few feet into the jungle on either side.

'It's like looking for a needle in a haystack,' Hal said. 'We don't stand one chance in a million.'

'What makes you think the stuff is here at all?' Roger was getting a little fed up with fighting brambles and thorns and the spiked edges of palm leaf stems.

'Just because I don't see where else they could take it. If there has been any ship around here we would have seen it. But somehow, the smugglers must have been watching us. When we succeeded in locating the wreck, that made it easy for them. They set out to grab as much of the booty as possible, hide it on the island, and after we leave they'll bring a ship to carry it away.'

They came out on the lagoon. It was circled by a lovely sand beach, now almost covered by the rising tide. In many places there was not room to walk between the water and the tree roots, and wading through the shallows slowed them up.

Thinking that perhaps Blake had passed this way, they looked for his trail on the submerged beach, but soon gave it up; the ripples of the rising tide would have wiped out all footprints, had there been any.

It was a full hour before they completed the circuit of the lagoon, and another hour before they could rejoin Skink under the breadfruit tree.

They were surprised not to find Dr Blake.

'That's strange,' Hal worried. 'He should have been here long ago. Something must have happened to him.'

'Now what could happen to him?' scoffed Skink.

'I don't know. Perhaps a broken ankle.'

No word could have made more of an impression upon Skink. The vivid picture of Blake's ankle caught in the jaws of the giant clam, and of Blake's vain effort to cut himself loose, made him shiver.

Hal eyed Skink closely. He noticed his trembling fingers, flushed cheeks, and feverish eyes. A walk in the sun could not do this. A horrible suspicion troubled him. He stooped suddenly and jerked Skink's knife from its sheath.

'What the devil are you doing?' complained Skink.

'I just want to see this knife.'

'Well, why not?' said Skink indifferently. 'But you could have asked for it, couldn't you?'

Hal studied the knife. Of course Skink would have cleaned it, but it was likely that some trace of blood would be left in the bevelling of the blade or in the grooved design of the handle. He searched carefully but could find nothing. He tossed the knife back to Skink.

'If I find there's been any crooked work,' he began grimly . . .

'Oh don't be so melodramatic, Hunt,' Skink cut in and, rising, started towards the dinghy. 'If you want to find Blake why don't you get to it instead of standing there making a stupid fool of yourself?'

The move took Hal by surprise. Skink had seemed reluctant to go in search of Blake, but now he was leading the way.

As for Skink, it had just penetrated his crime-crazed brain that, instead of trying to conceal

Blake's fate, he must reveal it. If Blake's body were not found they would believe that Skink had done away with him.

Now a new terror possessed his mind. They must hurry, hurry. Suppose the giant clam relaxed its hold. Suppose the body was carried away by the tide. Then Skink would be out of luck for he would have no evidence that Blake had not died a death of violence at his hands.

In the boat they rounded the island, closely hugging the shore. Now and then they shut off the motor and called. There was no answering call.

When they reached the bay of the disaster, Skink's mind was in confusion. How could he lead them to the spot without seeming to do so? It would be easy if he were at the tiller, but Hal was in the stern seat. Hal was still hugging the shore.

'Use your head, Hunt,' Skink said. 'He wouldn't climb that hill and down again. He'd swim across.'

Hal stubbornly held his course. 'There might have been a wide enough beach at the base of the cliff for him to walk around.'

But when he came to the foot of the cliff, he found the water so deep that even at low tide there could not have been any beach. Skink was right. Blake must have swum across. Perhaps he had drowned on the way, though why such a good swimmer as Blake should drown was a mystery, unless there had been foul play by Skink.

He circled to a spot where a crossing would naturally begin. Then he shut off the motor and told Roger to row, slowly.

In spite of the lack of evidence on Skink's blade,

he still half expected to come upon the drowned body of Blake with a knife wound between the shoulders.

He fished a mask out of a locker, put it on, and lowered his face into the water so that he could clearly see everything below.

The boat passed over a giant clam, its great jaws open. Then, ahead, he could make out another giant, its jaws closed upon some object, probably a large fish. Coming nearer he could see plainly what the object was, and his heart sank.

'Stop rowing,' he said to Roger. 'Here he is.'

He dived in, plunged his knife between the slightly separated edges of the shell and worked until he had made an opening large enough to admit his arm. He reached in and sank his knife into the powerful hinge. The huge valves eased apart.

Hal raised the limp body to the surface and the others helped lift it into the boat.

Hal climbed in and stripped off Blake's sodden shirt. There was no mark on the back or chest. The ankle was deeply cut. Hal thought he could see just what had happened.

'He was swimming across and got caught in the clam. He tried to saw off his foot but a knife isn't much good for that purpose. Before he could finish the tide rose and drowned him.'

One thing was as clear as the sun. Blake's death had been an accident. Skink was innocent. A fellow of idle boasts and mean tricks, but no murderer. Hal was glad, for he had never wanted to think the worst of Skink.

The three sat silent, each engaged in his own unhappy thoughts, while the dinghy bore its mournful burden to the schooner.

13
Burial Beneath the Sea

Blake had loved the colourful lands beneath the sea. He had spent much of his life studying their mysteries. Twice he had expressed the wish to be buried, like the seaman in the Jules Verne story, amid the loveliness and peace of the coral gardens.

His wish was respected. Hal and Roger selected the spot.

In a coral garden of surpassing beauty not far from the wreck of the *Santa Cruz* they came upon a splendid elkhorn coral in the form of a cross. Its erect column stood fifteen feet high and its two arms spanned five feet.

But not only in its great size was it superior to the ordinary graveyard cross. It was not built of dead granite or marble. It was a living and growing cross, the work of millions of Blake's small friends, the coral architects.

Its surface seemed inlaid with countless jewels of every colour that glowed softly in the light of the sun reaching down through ten fathoms of sea. It was a cross fit for the grave of a king — and the boys felt it was none too good for Blake. With pick-axe and shovel they dug a grave at the foot of the cross.

Returning to the deck, they joined in the service for burial at sea conducted by Captain Ike.

Then the body of the scientist, wrapped first in sailcloth and then in the flag of his country, was lowered over the side. Five pall-bearers, including Captain Ike and Omo who had insisted upon coming along although this was the captain's first experience with an aqualung, bore the shrouded form down into the depths.

Perhaps there had never been so strange a funeral procession as this. The grotesque masked and tanked figures that looked as if they might have come from Mars proceeded, head downward, pushing towards the bottom by thrusts of their enormous webbed feet.

Reaching the ocean floor, they walked with slow steps through an undersea paradise of great chrysanthemum-like anemones, stately fans and plumes, clouds of tiny rainbow-coloured fish, to the foot of the jewelled cross.

Reverently they laid the lover of the sea in his coral tomb, filled the grave with pure white sand, and paved it securely with masses of coral.

There were even flowers on the grave, for in the holes of the coral blocks brilliant sea anemones and gorgonias flourished.

These were flowers that would never fade, that would constantly renew themselves through the years and the centuries.

And so, under this carpet of life at the foot of the living cross in a garden that no building would ever violate, they left their friend to his long rest.

14
Kidnapped

Sadly the mourners returned to the deck of the *Lively Lady*.

But they could not sit and grieve. There was work to be done. Omo had made frequent inspections of the wreck during the day; now watches must be set for the night.

'You take the first, Roger, while there's still a little light,' Hal directed. 'Then I'll take an hour, then Omo. Then you again. Tomorrow we'll begin bringing up the cargo.'

'And who are you to be giving orders?' inquired Skink coolly.

Hal was surprised. 'Who else? You don't think *you* . . .'

'Have you forgotten that I was Blake's second in command?'

'He never said so.'

'Perhaps not in so many words. But didn't he bring me on because I was an experienced diver and you weren't? Didn't he give me the job of teaching you and your kid brother how to dive with the aqualung?'

Hal faced him angrily. 'That was before he found

out you were a crook and a coward. Then he put you down to leave on the next plane. And that still holds.'

Skink smiled with tolerant insolence. 'I'm sorry. I've changed my plans. I'm staying right here and you're going to take my orders.' He heard a snort from Captain Ike and turned upon him viciously. 'And so are you, you rickety old bag of leather and bones!'

A long arm attached to the bag of leather and bones began to swing, and when the open palm slapped Skink's face the force was sufficient to knock him clear across the deck in a heap under the gunwale.

'Mutiny! Mutiny!' screamed Skink. 'By the Holy Harry, I'll show you who's master here!'

He leaped below and came up at once with a revolver.

'Now, line up against that rail. I'll give each of you just one second to say who's boss. If you can't decide in that time I'll put you where you won't have to decide anything any more. Get going! Line up!' He brandished the revolver.

There was no rush to the rail. Instead, Hal began to move towards Skink.

'Get back!' yelled Skink, hopping up and down like a madman while his revolver wobbled wildly. 'Get back or I'll slug you!'

'Careful, Hunt,' advised Captain Ike. 'He's gone crazy. He's apt to do anything.'

'I don't think so,' Hal said. 'He hasn't the nerve to shoot. When he kills he goes round about — with a side-winder in a pocket, a scorpion in a helmet, a stonefish to do his killing for him . . .' He stopped and stared at Captain Ike. 'Or a giant clam!'

It came to him like a blinding flash of light. He did not know how, but somehow Skink had used the giant clam to accomplish the death of Dr Blake. It was exactly the sort of thing he would do — the sort of thing he had done with the sidewinder, the scorpion, and the stonefish. And when he had failed to warn Roger of the shark, when he had pretended the winch was out of order — it was all in the same pattern. His mind could not move except through sneaking, underhanded trickery. He lacked the courage to do anything straight out. He would not shoot.

Hal moved closer.

'One more step and you get it!' screamed Skink. His face was black with fury and his eyes bulged.

Hal made not only one more step, but a swift half-dozen. He struck the revolver out of Skink's hand and it flew over the gunwale into the sea. He gripped Skink by the throat and bore him down to the deck.

But Skink was as muscular and slippery as an eel. He slid out from under, leaped up, and kicked Hal in the face — or where Hal's face would have been had he not lifted it just at the right moment so that his opponent kicked an iron stanchion instead. Skink howled with pain.

The general laughter made him more furious. He pulled off his weighted belt. It was loaded with six lead discs each weighing a pound.

He swung the belt with all his force at Hal who retreated behind a mast. The belt whipped around the mast, the end of it slapping back towards Skink, and two pounds of solid metal caught him squarely on the side of the head, nearly knocking him out.

First the stanchion and now the mast – it seemed to the onlookers that the ship itself was fighting Skink. The *Lively Lady* was up in arms against him.

He tore the boom crutch from under the boom. It was a scissor-like support of very heavy wood, designed to keep the boom from swinging. Skink leaped up on to the rail in order to bring this weapon down like a club upon Hal's head.

A fresh breeze was blowing and at this instant the heavy boom swung to leeward. It swiped Skink from the rail and dropped him into the sea.

The *Lively Lady* had had the final word. She seemed to have said, 'Get off my clean decks and never come back.'

Skink put his hand on the ladder. Then he heard Hal's warning voice:

'If you come back on this ship you will be put in irons and held for the murder of Dr Blake.'

'Don't be ridiculous . . .' began Skink.

But when he glanced up at the row of angry faces looking down at him over the rail he knew it was useless to go on. His shipmates and his ship didn't want him. He had fooled them for the last time.

Well, almost the last time. He looked towards the island. It was a mile away, an easy pull for a good swimmer. He turned his back on the *Lively Lady* and struck out.

Hal was distressed and looked to Captain Ike for counsel.

'Should we have held him? We could overtake him in the dinghy and bring him back.'

Captain Ike shook his head.

'Let him go, lad, and good riddance. You couldn't have proved anything against him in court. There was no witness. There was no evidence that he laid a finger on Blake. No, you'll have to leave his punishment to the sky and the sea. And if I'm not mistaken . . .' he peered at a rolling formation of white and black clouds in the west, 'the sea and sky are getting ready to punish somebody.'

The expected turn in the weather did not come at once; the sea remained calm and the sky clear during the night. Roger took the first undersea watch, thankful that the ocean depths were not yet completely dark. After what had happened to Blake, his nerves all leaped up on end like a porcupine's quills every time a big fish approached or a strange sound struck his ears.

For a time he hovered like a helicopter above the deck of the *Santa Cruz*. Then, for something to do, he swam down into the hold and turned on his torch.

At once he saw that there had been a change. Several of the chests of treasure had disappeared.

It could not be Skink's doing this time because Skink had been on the island, and then with the funeral procession.

Anyone who didn't know Omo might suspect him, for he had been left alone to guard the wreck during the day. But Roger would as soon suspect his own brother as distrust their Polynesian friend.

At some time when Omo was not watching, the invisible man had made off with more of the loot.

Omo could hardly be blamed for this. He had made frequent visits to the wreck, and that was all that could be expected of him. No man could stay down continually because of the injurious effect of water pressure upon the body. The unseen smuggler had watched his chance and crept in while the wreck lay unprotected.

Certainly he would come back for more. He might even come back at this moment while no one was in sight and the ship appeared to be unguarded.

In a panic, Roger shot up out of the hold into the open and tried to look dangerous. He did not feel dangerous — only scared. The dusk was deepening and the moving objects in the sea were indistinct blobs. They could be fish or they might all be smugglers, no matter how he strained his eyes he could not be sure which. His hour seemed five hours long, but at last he was relieved by Hal. Before surfacing he took Hal into the hold and showed him the gaps in the cargo.

He was surprised to have Hal accompany him to the surface and on board the *Lively Lady*.

Hal tore the breathing tube from his mouth and gave vent to his anger.

'Omo! Get up out of that bunk! Captain Ike! No more sleep tonight! We're going to start salvaging, right now.'

Omo and the captain crawled out, blinking. 'Night is no time . . .' began the captain.

'They're looting the cargo. We won't give them a chance to take one more doubloon. The small stuff we can bring up in baskets and buckets. The big stuff the Iron Man can handle. Rig him up.'

Everyone hopped to with a will. Captain Ike took charge of deck operations while the others prepared to go below.

The Iron Man rose out of the hold, received Hal as its passenger, and sank into the sea. Two powerful searchlights attached to the chest of the monster were turned on. Hal gave instructions over the telephone, and Captain Ike regulated the cargo boom until the Iron Man was in position to sink through the open hatch into the hold. There the great arms embraced a huge chest and the signal was given to hoist away.

In the meantime Roger and Omo made a selection from the various baskets, buckets and nets that the ship could offer and descended to fill these containers with small articles. Up and down they went, steadily transferring the treasure of the *Santa Cruz* to the hold of the *Lively Lady*.

Hour after hour, with occasional rests, and intervals for recharging the tanks, the work went on.

Shortly before midnight Roger found himself alone in the hold. Hal and the Iron Man had gone up to deliver a bronze urn, and Omo had followed a moment later to recharge his aqualung.

Roger was occupied in filling a close-meshed net with gold bullion when he felt a tap on his shoulder.

Had Omo returned — or was it the touch of a fish or the tentacle of an octopus?

He swung the beam of his light around and up full into the faces of two masked men. He sprang to his feet and clapped his hand to his side, but his knife had been stolen from its sheath.

He struck out with his fists and had the satisfaction of knocking the aqualung mouthpiece from one man's mouth so that he must choke and swallow some water before he could recover it. Then he felt his arms gripped firmly and he was pushed up out of the hold and carried away swiftly, his kidnappers, one on either side of him, swimming with powerful strokes of their rubber fins.

Around them sparkled the million lights of the deepsea fish that had come to feed in the upper waters during the night. The glow faintly illuminated the coral gardens and the lone cross over the scientist's grave. Then they passed over the labyrinth of rocks.

A little beyond it they came down on the ocean floor beside what appeared to be an enormous boulder. But when the light of the torches picked it up he realized that it was a Japanese submarine like those he had seen manoeuvring in Truk lagoon.

On the near side was the bulge of the escape chamber. One of the men pulled open the trapdoor and Roger was pushed inside, the door closing behind him.

He heard the rushing sound as the water was forced out of the chamber by an intake of air. Then the trapdoor opened under his feet and he tumbled into the hold of the submarine.

Automatically the trapdoor closed and again he heard the rushing sound as the air in the escape chamber was displaced by water. Then the process was reversed and presently the inside trapdoor opened to spill one of his captors into the hold. The other soon followed.

They spat out their mouthpieces and whipped off their masks to reveal faces that Roger didn't like to look at. If his fate depended upon these thugs, his luck was out for sure. Their faces were twisted into permanent scowls and their eyes were as unfriendly as a moray eel's.

But apparently they were familiar with submarines — at least they began to push and pull a dozen gadgets as if they knew what they were about. They were too intent upon making a quick getaway to pay any attention to Roger. There was the whoosh of water from the ballast tanks to give the craft positive buoyancy, rising it from the bottom, and the purr of an electric motor, to start the propellor. One man sat at the wheel with his eye on the compass while the other watched the fathometer, showing the clearance between the submarine and the bottom.

Presently the deck under Roger's feet tilted upward more steeply as if the submarine were coming to the surface and the man who was steering glued his eye to the periscope. After a time the motor stopped, an overhead hatch was opened and the fresh night air flooded in.

One of the men, the one with an ugly scar over

his left eye, said in a rasping voice. 'All right, buddy. End of the line.'

Roger climbed up through the hatch. The men followed, lugging a heavy chest, evidently stolen from the wreck.

'Just hop off and swim ashore,' instructed Scarface. 'Reception committee waiting for you on the beach.'

Roger swam and waded ashore. A dark form stood on the beach. Roger heard a low laugh – it was Skink's laugh.

'So nice of you to join us,' Skink said. 'We haven't much to offer you, but you may be sure we'll do our best to make you uncomfortable.'

Scarface waded ashore.

'Did you leave my note?' Skink asked.

'Sure, boss. Tied it to the mast, just like you said.'

Roger's mind buzzed with questions but he refused to ask them, knowing he would not get honest answers.

'Now, if you'll just follow me,' said Skink, keeping up his show of mock courtesy, 'and pardon me for going first, but I happen to know the way.'

He struck off into the thicket, lighting his way with his torch. The two thugs stuck unpleasantly close to Roger and any thought of escaping into the jungle had to be dismissed.

For fifteen or twenty minutes they wormed their way through the brush, then came out before a tent in a small clearing.

'Be it ever so humble,' said Skink, 'it's home, sweet home. Make a fire, Chubb. Time for a midnight snack before you get back to the wreck.'

Roger, with a boy's eternal interest in food, pricked up his ears at mention of a snack.

'Of course our guest here won't want anything,' Skink went on. 'It's not good to eat when you're nervous.' He played his light in Roger's face. 'You are nervous aren't you?'

Roger could take no more.

'Just this nervous,' he said, and let fly with his fists. Skink, taken by surprise, went down like a ninepin. Roger was on him in an instant, pummelling that smug face.

The two men dragged him off and bashed his head against a tree trunk until he faded out. They left him in an unconscious heap and began their midnight supper.

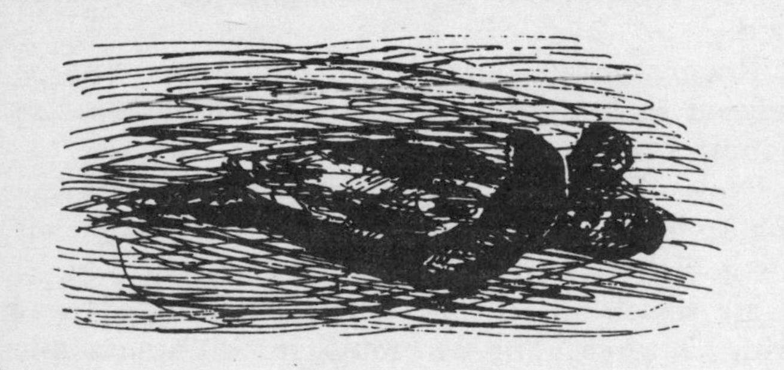

15
Battle of the Sea Bottom

The Iron Man sank into the hold of the *Santa Cruz*. Hal, peering through the quartz eyes of the metal monster, was surprised to see that Roger was no longer working where he had seen him last.

Had the boy wearied of his labours and gone exploring?

'I don't see Roger,' Hal telephoned to Captain Ike. 'Tell Omo to get down here as soon as possible and look for him.'

It was another five minutes before Omo could finish recharging his aqualung and reach the wreck. He systematically searched the hold. Then he went up on deck and explored the two castles. He swam a few yards away from the ship and made a complete circuit around it. Then he rose to the surface and reported to Captain Ike. The captain telephoned Hal.

'Omo has been all through the wreck and around it. He can't find your brother.'

'Haul me up,' Hal said.

The Iron Man rose, clutching a marble statue of

Venus that had once perhaps graced the governor's garden. The black monster and the white goddess in a fond embrace broke the surface, soared into the air, and came down upon the deck. 'Let me out,' Hal ordered. The trapdoor was unbolted and he crawled out, immediately calling for his aqualung and mask.

'Let's go down and take another look.'

They explored the wreck thoroughly, looking in every nook and cranny to be sure that a giant octopus had not drawn Roger into its hole, visiting the cross on the chance that Roger might have made a sentimental journey to the scientist's grave, even penetrating the labyrinth of rocks to the cave since Roger might have gone there to see if it was still being used as a transfer point for stolen treasure.

With heavy hearts they returned to the wreck. In the soft glow of sea life Hal noticed a black object tied to the top of one of the broken masts. He swam closer. It was a bottle. He tore it loose, signalled to Omo, and they ascended to the *Lively Lady*.

In feverish haste, Hal knocked off the head of the bottle. Inside was a paper. He fished it out and spread it open under the beam of his torch. Hal recognized Skink's writing.

HUNT:

Your brother is being held for ransom. We demand half a million dollars for his release. We make it easy for you to meet these terms. All you need do is return to Truk and leave the wreck of the *Santa Cruz* to us. Give us one week to remove the treasure. At the end of that time your brother will be delivered to you, unharmed, at Truk.

S. K. INKHAM

The three sat in stunned silence. Hal's first impulse was to abandon the wreck and return to Truk. He must let Skink have his way. Anything to save his brother. Captain Ike's and Omo's thoughts were running in the same direction.

'Skink wins,' the captain said. 'He's outsmarted us. I always said he was a sly one. Shall I up anchor and make for Truk?'

'I don't see what else we can do,' Omo said.

But Hal's mind had taken a new tack. Was he going to let himself be outsmarted by Skink? And how about his job? It was easy for the captain and Omo to talk about sailing away. Their duty was to the ship; but his duty was to the Institute. Dr Blake had been instructed to salvage the cargo of the *Santa Cruz* and now that Dr Blake was no longer here that responsibility was Hal's.

'We have a job here,' he said '— to bring up that treasure. We can't let ourselves be scared off by a letter from a pack of bandits.'

'But how about Roger?' Omo asked.

'It's Roger's job too. He wouldn't want us to shirk it just to save his skin. He would be humiliated to think that the whole expedition had been wrecked because of him. I know he wouldn't want it that way. Let's get on with our job. They won't be expecting that, and perhaps we can get a lot of the stuff up before they interfere with us. If they do interfere, we'll give them a fight that they'll remember.'

The two divers again descended to the wreck, Omo with his aqualung, Hal operating the Iron Man. They worked energetically and the store of treasure in the hold of the *Lively Lady* grew steadily.

Yet their nerves were on edge because they knew this could not go on without interruption. Something would happen – what it would be they could hardly guess.

Each time they came up they found Captain Ike a little more anxious about the weather. There was every sign of an approaching storm. The barometer had slid from 30 to 29.3 and was still on the way down. But Hal would not consider stopping the work to run for shelter.

It was about two hours after midnight when, as the Iron Man sank towards the wreck, the search-lights picked up a roundish object that Hal at first took to be a whale. As it came closer he saw that it was one of the Truk submarines.

It seemed to be making straight for him and he called over the telephone. 'Haul up, quick!'

Before the Iron Man could begin to rise it was struck a terrific blow on the right side by the nose of the submarine. The crash sent Hal hurling against the steel wall and a shower of broken instruments fell upon him. He called to Captain Ike but got no answer – the wire had evidently been snapped. The impact started the cable spinning off the winch drum and the Iron Man sank to the bottom. There it settled on its side. The lights had gone out and water was leaking in.

Again there was a tremendous crash as the submarine renewed the attack and Hal was battered and bruised in the steel shell. The submarine's light cast weird beams through the water as it circled and came down beside the Iron Man.

A trapdoor opened and a figure came out. It swam to the rear port of the Iron Man and seemed to be trying to unbolt it. There was a sudden inrush of water as the trapdoor opened and Hal felt his body contract under the pressure.

He wore no aqualung, since that was not required inside the Iron Man, and would certainly drown if he could not immediately escape to the surface.

He started to crawl out through the trapdoor and was surprised to find somebody helping him. He looked up into a face that he knew was Skink's, in spite of the fact that it was half covered by mask and mouthpiece.

Skink was trying to haul him over to the submarine. Hal, weakened by his battering inside the diving bell, was still strong enough to fight back vigorously.

His first blow knocked the aqualung tube from Skink's mouth and every time Skink replaced it he knocked it out again. He had no air, and Skink should have none either. They could last for perhaps two minutes, three at the outside — then they would drown together.

Locked in each other's grip, they wrestled through a grove of elkhorn coral. Hal got his hands around Skink's neck and choked him until his eyes

bulged. Then he threw him upon a bed of fire coral, the most poisonous coral of the sea.

At last he was free and began to shoot towards the surface, but was caught by the leg and hauled down to meet two tough blackguards who immediately jammed him into the escape chamber of the sub and closed the trapdoor.

The water drained away and suddenly he could breathe again. The trapdoor beneath him opened and he dropped into the submarine.

A few minutes later the senseless body of Skink dropped beside him. The two other men followed. The chamber was too small for four men, but Skink and Hal, both completely exhausted, were thrown on shelves like sacks of potatoes, while the other two navigated the undersea boat to the island.

Arriving at the beach, the men hoisted Skink's still unconscious body through the hatch to the deck. The cool air revived him and he was able, with some help, to swim ashore and stagger through the jungle to the bandits' camp.

'Well, boss,' chuckled one of his henchmen, 'the kid sure gave you a drubbing.'

'That's nothing to what I'll do to him,' growled Skink. When he reached camp he was not in a condition to do anything to anybody. He dropped in a heap and began scratching himself furiously. Red welts were breaking out all over his body, thanks to the fire coral.

Hal looked about anxiously for his brother. 'Roger!' he called. The sickening fear came over him that these devils had already done away with the boy. He drew open the flap of the tent.

Roger lay on the ground trussed up like a chicken, tied hand and foot, a gag in his mouth. But his eyes were open and bright, blinking in the light of the torches. Hal jerked the gag from his mouth.

His lips and tongue were swollen and cramped by the gagging, but he managed to say, 'Gee, am I glad to see you!' He noticed that the two men flanking Hal had caught his arms in a firm grip. 'I see you've already met my friends. This is Chubb and this is Scarface.'

The latter evidently didn't like the name Roger had given him. 'I'll kick that sense of humour out of you,' he snarled, and swung his brawny foot into Roger's ribs.

Hal wrenched an arm loose and landed a hook to the ruffian's jaw. At once there was a wild scrimmage in which even Skink came to take part. The three men crashed Hal to the ground, tied him hand and foot, and gagged him. Roger too was gagged. But Skink was not satisfied.

'I think we ought to finish them both off. Chubb, put some bullets into them.'

'Now looky here,' complained Chubb. 'If you want anything like that done, do it yourself. We're not looking for any murder rap. We're apt to get into enough trouble as it is . . .'

Skink cut him off. 'I employed you and you'll do as I say.'

Chubb loomed over him, clenching his fists. 'Don't forget we have some say in this thing too, you miserable little shrimp. You couldn't do without us. We stole the sub for you, didn't we, and we know how to handle it, thanks to ten years in the sub service.'

'From which you were dishonourably discharged,' taunted Skink. 'You've been fired from the navy and you've stolen a sub. If you bop off these two chumps how can that put you in any worse than you are already?'

'Again I ask,' insisted Chubb, 'why don't you do it yourself?'

Skink was about to answer when a swirl of wind swept through the trees with a sound of breaking branches. The tent leaped from the ground, tore itself loose from its guy ropes, and wrapped itself around a palm trunk. Far through the jungle came a booming, droning, shrilling sound as of a great orchestra tuning up.

Scarface stared at the sky. 'Typhoon!' he exclaimed.

Palm fronds went sailing among the stars. Heavy coconuts thudded to the ground. A dead tree crashed within a few inches of the spot where Hal and Roger lay.

In the moment's lull that followed, Skink said, 'None of us is going to have to do any killing. Mother Nature will do it for us. Leave them right there – Mother will take care of them.'

Chubb looked around in pop-eyed fear. 'But how about us? This part of the island is low. The sea will flood it.'

The wind came again, louder and stronger.

'Lucky you have me to think for you,' cried Skink. 'Quick – into the sub and submerge.' He struck off towards the beach. 'Ten fathoms down we won't even know there *is* a storm.'

Hal and Roger watched them go, watched until the last flicker of their torches had been swallowed by the jungle. The trees were tossing in a frenzy. Branches and nuts rained to the ground. Above the scream of the wind came the pound of the surf on the windward shore. That was the worst danger – a rising sea, a drowned island.

Hal hunched himself over to his brother's side, felt about until he got his fingers on the ropes that bound him. He began picking with his own bound hands at the hard knots.

16
The Typhoon

A flash of lightning made the clearing as bright as day. When it was gone the night seemed darker than ever. A terrific explosion of thunder rose for a moment above the howl of the wind.

Then came the rain. Not in drops, but in chunks, blocks, tons. It was as if there had been a lake in the sky and the bottom had fallen out of it.

A typhoon does not simply release water from the clouds. It hurls it down angrily as if determined to smash to a pulp everything on the face of the earth. The bombardment beat upon the flesh like a hail of stones.

Hal, his hands tied behind him, made slow work of freeing Roger. The water did not make it any easier to untie the knots. It was well on towards dawn before he succeeded.

Then Roger tore the gags from his brother's mouth and his own, and went to work on Hal's wrists. When the ankles too were freed, dawn had definitely arrived.

But what a dawn! One could almost wish for the dark again. The sky was one great boiling mudpot of black clouds, split now and then by neon-bright flashes of lightning.

Near the horizon the black faded into a sickly white and when the sun rose it looked half its usual size and as brown as a Polynesian face.

Most of the trees were broken off part way up. The tall stumps that remained vibrated like tuning forks. Every moment another tree would give up its struggle, there would be a crack like the report of a pistol, and the head of a palm would snap off and sail away. It would probably keep on sailing far out into the ocean. Sometimes instead of sailing horizontally the palm heads would be carried up into the sky and disappear into the black clouds.

The roar of the thunder, the crash of the rain, the scream and shriek of the wind, the breaking of limbs and fall of trees — it was almost more than the ears and nerves could stand.

Hal stood up but was immediately thrown to the earth again by wind and rain. Perhaps if they lay low for a while there would be a lull. Neither spoke — even a shout would not have been heard above that din. They huddled in balls and tried to present as little surface as possible to the storm.

Then came the sea. Hal saw the tongues of water

licking over the ground. They looked so innocent, but they meant death. He dipped his finger in one of them and touched his tongue to be sure it was not rain water. It was salt. The ocean was preparing to swallow the island.

They must get to high ground. He signalled to Roger and began to crawl on hands and knees.

It was necessary to keep a sharp eye out for coconuts. They no longer fell. They flew sideways as if shot from guns. Branches, sticks, leaves fled furiously past. Now and then a tree would be ripped up bodily by the roots and tear a path for itself straight through the underbrush and out to sea.

The crash of the surf on the beach sent a tremor through the island that could be distinctly felt. Regularly it came, like an overstrong heartbeat.

But suddenly there was an entirely new convulsion. The earth shook in the grip of a violent earthquake accompanied by a roar louder than the voice of the thunder.

Presently the rain stopped. That was a relief. But the wind was hot and now, no longer cooled by the rain, it struck the skin like the heat of a blast furnace when the door is suddenly thrown open.

The tongues of salt water had become rushing creeks. Still keeping their hands on the ground to avoid the full force of the wind, the boys sometimes found it hard to keep their faces above water.

Not all the water was coming across the land. Some was coming up through it. Whether by way of caverns beneath the island, or through the soil itself, the stupendous pressure of the storm forced liquid into solid and made the whole thing half liquid. If one did not drown in the rising sea, there was the danger of sinking in this newly formed quicksand.

Roger seized Hal's arm and pointed up. A gigantic comber that seemed fifty feet high was roaring towards them. Its curling crest was loaded with bunches of coconuts, branches, brush and whole trees. It was a terrific picture of majesty and power, and its thunder drowned the shriek of the wind.

Hal yanked Roger towards a tamanu tree, the sturdiest thing that grows on a South Sea island. In feverish haste they clambered up through its great branches. Before they could reach the top the comber struck.

The tree shivered. Huge boulders crashed into the branches and trunk. The logs and trees carried by the wave caught on the limbs and did not reach the two terrified climbers.

But they could not climb high enough to escape that roaring crest. As it struck them they instinctively closed their eyes, gritted their teeth, held their breath.

The comber plucked them from the tree as if they had been leaves. It pitched them this way and that through the branches and then out into space. It pummelled them with sticks and stones. It carried them high and away over the island and finally flung them, bleeding and almost unconscious, against a mass of branches.

The great wave passed on leaving comparative stillness in its wake. The backwash drained the water from the land. For a few moments it would be almost dry. The super-waves in a typhoon were usually about a quarter of a mile apart.

If they could get to high ground before the next came along . . .

'Now's our chance. Come on.'

Hal pulled Roger up and they plunged into the wind. On hands and knees they fought it. It was like something solid. They had to bore a hole in it, tunnel through it. On every side it was ripping the island to bits. No wonder the Polynesians called the typhoon, 'The Wind that Overturns the Land'.

Another earthquake shook the island. This one was worse than the first. It opened cracks in the ground six feet wide and scores of trees weakened by the wind crashed down.

Now they were on rising ground. They climbed a steep hill and when the next comber arrived it washed by below them. They came out on the edge of a cliff overlooking the sea.

Hal at once recognized the spot. Down there in the mouth of the bay Dr Blake had died.

The bay was directly in the path of the storm. Through its wide mouth the waves came rolling in, towering higher as they reached the shallows, turning the whole bay into an inferno of clashing waters, finally crashing into the cliff and sending spray two hundred feet high.

The boys looked anxiously for the *Lively Lady*. She was no longer at her customary moorings. They scanned the western sea in vain.

Then they turned and looked into the lagoon. And there was the *Lively Lady* — climbing a hill!

Wisely, Captain Ike had sought the shelter of the

lagoon. It was poor shelter, though better than none. The wind swept furiously across it and its own small waves were joined by the waves of the sea which raged over the low parts of the island and then over the lagoon.

Wind and wave had pushed the *Lively Lady* ashore on a sandy slope and every new wave carried it a little higher. It was already a good thirty feet above the normal level of the lagoon. The boys gazed in astonishment at the incredible spectacle of a ship climbing a hill.

Not that she was doing it very gallantly, for she was on her beam ends. Both masts were broken and holes gaped in her hull.

But she was climbing to safety. Safety, unless she were washed clear over the hill and into the sea beyond.

Every wave lifted her a few inches, and each of the giant combers carried her six or ten feet higher. The boys could only hope that Captain Ike and Omo were still in the ship, and alive, and would survive this terrible battering.

And what of the other three men who had taken refuge in the submarine? Skink had been shrewd to think of that. Down in the ocean depths, their chances should be best of all.

Hal knew that the miniature submarines were not equipped to go deeper than ten fathoms, but that should be deep enough. All would be quiet and peaceful in the coral gardens.

And yet he knew that this was not necessarily true. Even at a depth of hundreds of fathoms scientists had encountered and measured undersea waves of tremendous force. Violent currents, rivers, torrents, had been found in the depths. And in a typhoon, who could say what would happen?

Something strange was happening right now. Every few moments the ground shook under some terrific impact. Hal crawled to the edge of the cliff and looked over. He was just in time to see an enormous boulder at least a dozen feet in diameter flung bodily against the face of the cliff. The terrific collision caused it to break into fragments which fell back into the sea.

Presently one of the giant combers rushed in to hurl three more colossal rocks against the face of the cliff.

But where did they come from? There had been no such boulders in the bay. They must come from the open sea. Then he remembered the labyrinth of rocks near the wreck of the *Santa Cruz*.

Terrific forces were at work in those depths, violent undertows and upwellings that rolled the rocks towards shore until they came into water shallow enough for the combers to get a grip on them and carry them to the cliff.

He had a moment's dread that the same thing would happen to the wreck of the *Santa Cruz*. It might be torn loose, broken to bits, and all its treasures scattered. But he realized that this was not likely to happen. The wreck was so deeply buried in sand that the typhoons of three hundred years had failed to budge it.

Roger, lying beside him, his face stung with shafts of spray, was staring fixedly out to sea. Hal followed his gaze and saw something that made his heart sink. It came surging towards the bay in the grip of a giant comber. The *Santa Cruz* had been ripped from the sea bottom after all and was being carried to destruction.

No, it was not the *Santa Cruz*. It was black like one of the great boulders, but it was not a boulder.

Then Roger screamed into his ear. 'The submarine!'

It was the submarine. The angry upwelling sea had thrown it up in spite of all its efforts to stay deep and safe. The god of the sea was about to deliver the three criminals to a terrible justice. The submarine looked like a black bubble, so lightly was it carried in the claws of the great wave.

In through the mouth of the bay it was swept and over the scene of the scientist's tragic death. The comber seemed to lift it with mighty arms as the shallows forced the wave higher. The black bubble spun dizzily round and round. The terror of the three men inside it must be beyond all imagining.

Then it struck. It shook the precipice and parts of the rock face fell away. The black thing exploded like a bomb. Steel fragments flew in all directions, some barely missing the faces of the two horror-stricken boys who looked down over the edge of the cliff. The bodies of men were seen indistinctly through the spray as they were tossed in high arcs and fell back into the boiling sea.

Then came the tremendous backwash with a sound like a profound sigh as if the sea was deeply satisfied with what it had done.

Skink, who had thought it smart to make Nature

do his dirty work for him, whether by serpent, scorpion, stonefish or giant clam, who had contrived Blake's death without laying a hand on him, who had left two victims bound and gagged at the mercy of the typhoon while he and his pals sought safety in the arms of Ocean – Skink had now been outsmarted by Nature herself.

Hal could not be happy about it. His head ached and he noticed that his brother's face was as green as turtle soup. Another earthquake shook the island and parts of the precipice fell away. The boys warily retreated to a safer position.

The great combers were fewer now and less violent. The wind was no longer a solid wall. It had begun to flutter and veer, as if uncertain what to do next.

For an hour it grew more and more nervous, then suddenly shot away, leaving a dead calm behind it.

It could be heard in the distance, scurrying off in search of new lands to lay waste. The waters still dashed below but they no longer had push and purpose.

The spray cleared and the island could be seen in all its bleak desolation. It had been chewed down to a third of its former size. If the typhoon had continued a day longer the island must have disappeared completely.

Remaining were two hills and a low stretch covered with a dozen feet of water through which projected hundreds of broken stumps. Nowhere was there one whole tree. Boulders as big as cottages lined the windward shore.

By afternoon the waves had gone down and the water that flooded the lowlands was subsiding. And there, coming across the island, was a boat!

It was the dinghy of the *Lively Lady* and there were two figures in it. Hal and Roger shouted for joy and waved and got an answering signal.

It was a wonderful reunion as the dinghy pulled through between two tree stumps and beached on the slope of the hill.

'And how's the *Lively Lady*?' Hal asked.

'Pretty well smashed up,' Captain Ike answered, 'but she can be fixed.'

'Our sea serpent and electric ray and moray and the rest — are they safe?'

'They should be all right. When the blow started I filled the tanks to the brim, then sealed them shut so there would be no sloshing. I don't suppose the specimens in those tanks got as much of a beating as we got outside.'

'It must have been pretty rough, going up that grade.' Hal laughed at the sight of the schooner perched on top of the hill sixty feet above the lagoon. 'Just like the Ark on Mount Ararat,' he said.

The birds had all been blown away, but now one appeared in the distance. 'A big one,' Roger said. 'It must be a frigate bird.'

Captain Ike squinted hard. 'Better than that,' he said. 'It's a helicopter from the naval base.'

The plane circled the schooner on the hilltop, then crossed to the other hill and settled within a few feet of the marooned sailors. The pilot called down:

'Any passengers for Truk? Or do you like it here?'

They climbed up in a hurry as if afraid he would change his mind. The plane winged away to the north.

'We wondered how you were coming through the blow,' the pilot shouted over the roar of the engine. 'Thought we'd better come and see if there was anything left of you. But weren't there two more? Where are they?'

Hal told the story of Dr Blake and Skink, and told it again to the commanding officer at the base. Then came hot food, plenty of it, and clean sheets, and hard-earned sleep.

17
The Call of the Volcanoes

The rest of the story is quickly told.

With equipment borrowed from the naval base, the 'Ark' was brought down from its Ararat, towed to Truk and there reconditioned.

The cache of treasure on the island was discovered under a mat of brush and debris, and the operations of salvaging the cargo of the *Santa Cruz* continued until it had all been transferred to the *Lively Lady*, carried to Truk, and there loaded on a cargo vessel bound for San Francisco. On the same ship, in specially built tanks, went the valuable specimens of deep-sea life.

The map of many island groups had been suddenly changed. The typhoon had wiped out dozens of small islands, and volcanic disturbances had thrown up islands where none had existed before. Several active volcanoes had poked their noses out of the sea and were spouting ashes and white-hot lava.

Earthquakes continued to be reported throughout the western Pacific, and volcanoes in Japan, Hawaii, the Philippines and Indonesia blazed into activity.

A volcanologist from the American Museum of Natural History, hearing that the Oceanographic Institute no longer needed the services of the *Lively Lady*, flew to Truk. He boarded the schooner.

'It's a fine ship,' he told Hal and Captain Ike. 'Just what we need. I want to visit some of these brand-new islands that are exploding to the surface – and the volcanoes. This whole end of the Pacific seems to be blowing up. Something very unusual is happening; we want to find out what. How about it – is your ship available?'

Captain Ike squinted about at Hal, Roger and Omo. They looked very unhappy. Captain Ike knew what that meant.

'I suppose so,' he said slowly, 'though we'd be mighty sorry to part with her.'

'Part with her!' exclaimed the visitor. 'Nothing was farther from my thought. I've had the finest reports about you from the Institute. I want every one of you to come with me. I couldn't ask for better assistants.'

Everyone brightened as if by magic. Hal spoke for them. 'That sounds good to us.'

The scientist raised a warning hand. 'Don't decide too quickly. It's dangerous business – going down at the end of a rope into an erupting crater.'

'If it's all right with you,' Hal said, looking about at his companions who were nodding in vigorous agreement, 'it's all right with us.'

And so the *Lively Lady* was commissioned for a strange voyage, the story of which will be found in another book of this series, *Volcano Adventure.*

ADVENTURE

The Adventure Series by Willard Price

Read these exciting stories about Hal and Roger Hunt and their search for wild animals. Out now in paperback from Red Fox at £3.99

Amazon Adventure

Hal and Roger find themselves abandoned and alone in the Amazon Jungle when a mission to explore unchartered territory of the Pastaza River goes off course...

0 09 918221 1

Underwater Adventure

The intrepid Hunts have joined forces with the Oceanographic Institute to study sea life, collect specimens and follow a sunken treasure ship trail...

0 09 918231 9

Arctic Adventure

Olrik the eskimo and his bear, Nanook, join Hal and Roger on their trek towards the polar ice cap. And with Zeb the hunter hot on their trail the temperature soon turns from cold to murderously chilling...

0 09 918321 8

Elephant Adventure

Danger levels soar with the temperature for Hal and Roger as they embark upon a journey to the equator, charged with the task of finding an extremely rare white elephant...

0 09 918331 5

Volcano Adventure

A scientific study of the volcanoes of the Pacific with world famous volcanologist, Dr Dan Adams, erupts into an adventure of a lifetime for Hal and Roger....

0 09 918241 6

South Sea Adventure

Hal and Roger can't resist the offer of a trip to the South Seas in search of a creature known as the Nightmare of the Pacific...

0 09 918251 3

Safari Adventure

Tsavo national park has become a death trap. Can Hal and Roger succeed in their mission of liberating it from the clutches of a Blackbeard's deadly gang of poachers?...

0 09 918341 2

African Adventure

On safari in African big-game country, Hal and Roger coolly tackle their brief to round up a mysterious man-eating beast. Meanwhile, a merciless band of killers follow in their wake...

0 09 918371 4

It's wild! It's dangerous! And it's out there!